Disaster Resilience and Green Growth

In recent years, it has become clear that environmental changes and disaster risks are closely linked. Climate change, changes in land use, and the depletion of natural resources make communities and infrastructure more vulnerable. This has created a need for comprehensive and collaborative solutions that bring together different fields. Effective management of disaster risks and adaptation to climate change requires coordinated efforts across infrastructure, social systems, and ecological resilience. Factors related to the environment and human activities greatly influence risk exposure. Therefore, risk management must consider these elements. Bridging fields such as environmental science, engineering, economics, and policy can lead to effective disaster resilience.

Traditional economic growth models often overlook environmental sustainability, resulting in more resource depletion and vulnerability. In response, new green growth strategies focus on sustainability, including nature-based solutions and climate actions to promote circular economies and low-carbon development, enhancing the approach of disaster resilience. Current policy discussions emphasize the importance of building resilient systems in infrastructure, industry, and public health as part of the sustainability agenda post-2030. Furthermore, new energy technologies present both opportunities and risks, requiring strategies to manage their impacts effectively. Chemical and industrial safety risks in addition to natural hazard induced technological disasters as complex emergencies are also a growing concerns in disaster management and sustainability in the emerging world.

The scope of book series relates all the SDG in implicit ways but is focused particularly on SDG 2, 3, 8, 9, 11, 12, 13 and 15. The book series "Disaster Resilience and Green Growth" aims to share research and insights on these topics surrounding safety and resilience of human life and society in the changing world, fostering innovation and knowledge-sharing for a sustainable future.

Joice K. Joseph • A. P. Pradeepkumar
Editors

Eco-Social Work for Disaster Risk Reduction

Advancing Social and Environmental Justice

Springer

Editors
Joice K. Joseph
Department of MSW Disaster Management
Loyola College of Social Sciences
(Autonomous)
Thiruvananthapuram, Kerala, India

A. P. Pradeepkumar
Department of Geology
University of Kerala
Thiruvananthapuram, Kerala, India

ISSN 2662-4885 ISSN 2662-4893 (electronic)
Disaster Resilience and Green Growth
ISBN 978-981-95-4884-2 ISBN 978-981-95-4885-9 (eBook)
https://doi.org/10.1007/978-981-95-4885-9

This Springer imprint is published by the registered company Springer Nature Singapore Pte Ltd.
The registered company address is: 152 Beach Road, #21-01/04 Gateway East, Singapore 189721, Singapore

Foreword

What a refreshing new addition this book is to our social work literature, capturing as it does the significance of eco-social work practice in the context of climate changes and community fragility. Across the world, social workers are realizing the significance of climate disasters to human health, resilience, and well-being, and are outlining examples of the fragility of ecosystems and communities. This book is no exception outlining as it does a call to arms for social workers to recognize the links between climate volatility, degraded ecosystems, and the vulnerability of affected communities. Drawing on several examples from India of climate crises, the book is firmly based on social work practice. It draws out the need for workers to ground their practice in eco-social work theory, incorporating this into not only the way we address the impacts of disasters on people, communities, and environments, but also how we address disaster governance and assist our communities to build resilience and sustainability.

The book widens our understanding of social work, disaster risk reduction, social justice, community resilience, and eco-social work and of the interdependence of humans and ecosystems. The provision of community examples such as Indigenous communities working in forest management in the hills of Kerala and fishing communities operating along the Indian coastline adds significant insights to our understanding of the links between ecological and social systems. The book deepens our understanding of resilience and provides yet another example of the significance of Indigenous knowledge to our understanding of ecosystem management.

As the various authors note, it is becoming increasingly obvious that social workers must incorporate into our professional standards and our practice not only support for the Sendai Framework and the accompanying gender and disaster policies but also our facilitation of the Sustainable Development Goals (SDGs) at local levels. Because we are deeply embedded in community, we are at the leading edge of assessing and building a response to SDG implementation. Further, and of critical importance, is that our location at community levels makes us critical experts in the reshaping of what comes next after the SDGs finish in 2030. This is where the book—*Eco-Social Work for Disaster Risk Reduction: Advancing Social and Environmental Justice*, edited by Dr. Joice K Joseph and Prof. A P Pradeepkumar, is so critical to the way we shape our response to disasters and ecosystem sustainability, to building resilience and to providing our expertise of community level impacts to global fora. They remind us that social work is not only an international profession, it is also a profession that has the capacity to lead the fight for social and environmental sustainability in a global context.

Professor of Social Work, University of Newcastle Margaret Alston
Callaghan, NSW, Australia
Emerita Professor, Monash University
Melbourne, VIC, Australia

Preface

This volume, *Eco-Social Work for Disaster Risk Reduction: Advancing Social and Environmental Justice*, foregrounds the urgent need for integrated responses to the converging crises of climate change, ecological degradation, and social vulnerability. The twenty-first century is witnessing an unprecedented convergence of ecological crises and social challenges. Climate change, desertification, biodiversity loss, pollution, forced migration, and energy insecurity are no longer distant threats but lived realities that cut across geographies and social strata. Their impacts are felt most acutely in disasters, which have grown in frequency, intensity, and complexity. According to the Centre for Research on the Epidemiology of Disasters (CRED), in 2024 alone 393 disaster events were recorded worldwide, causing over 16,700 deaths, affecting 167 million people, and inflicting an estimated US$241.95 billion in economic losses.

Disasters, however, are not "natural." They emerge from the intersection of hazards with social vulnerabilities, shaped by inequality, exclusion, unsustainable development, and environmental mismanagement. As Ilan Kelman of University College London reminds us, disasters are socially constructed, determined less by hazard intensity than by the decisions, systems, and structures that create and perpetuate risk. It is in this context that eco-social work is gaining salience as a transformative practice framework for Disaster Risk Reduction (DRR). Eco-social work offers precisely such a paradigm: one that is holistic, relational, and grounded in the interdependence of people, communities, and ecosystems. It builds on the evolution of environmental social work—from Jane Adams' insistence on addressing unsanitary and unsafe environments as central to social reform, to the radical ecological awakening spurred by Rachel Carson's *Silent Spring*, through to the Brundtland Commission, the Kyoto Protocol, and the urgent calls of the Intergovernmental Panel on Climate Change (IPCC) to be mindful of climate change.

Eco-social work represents a "third generation" of environmental social work. Building on the *person-in-environment* and *green social work* paradigms, it extends practice to macro contexts, where climate change, ecological degradation, and disaster risk converge. It emphasizes sustainability, justice, and inclusion as inseparable goals, highlighting the need for systemic responses that integrate environmental and social concerns. The *2012 Global Agenda for Social Work and Social Development* and the *2022 World Social Work Day* theme, *Co-building a New Eco-Social World: Leaving No One Behind*, further affirm the profession's responsibility to address ecological justice alongside human well-being. This book, *Eco-Social Work for Disaster Risk Reduction: Advancing Social and Environmental Justice*, responds to the growing demand for scholarship and practice that bridges social work, ecology, and disaster studies. It explores how eco-social principles can be applied to reduce disaster risks, enhance resilience, and promote sustainable development. This work highlights how eco-social work integrates theory, case studies, and practice-based evidence to address climate change, environmental degradation, and disaster risk governance through community-centered, inclusive, and rights-based approaches and further reframes vulnerability and risk through a justice lens, emphasizing social protection, sustainable livelihoods, and ecological stewardship to ensure that marginalized populations are not left behind. By bringing together interdisciplinary perspectives, the book highlights eco-social work as both a field of scholarship and a mode of praxis with transformative potential.

This book is designed for a wide readership; academics from social work, environmental studies, public policy, and DRR are the prime target audience for this book along with development practitioners, humanitarian actors, and policymakers. Its intent is not only to inform but also to inspire transformative action—encouraging professionals to recognize their role in addressing the intertwined challenges of environmental sustainability and social equity.

The volume comprises 12 chapters organized into thematic sections—foundations and frameworks, practice perspectives, emerging frontiers, and case studies—that together illustrate how eco-social work advances social and environmental justice in disaster contexts. These chapters examine the nexus of social work, environmental justice, and community resilience, underscoring the core argument that disasters are fundamentally socially manufactured events shaped by preexisting social, economic, and environmental inequalities. The research articles together emphasize that effective DRR has to reconcile the historical disconnect between environmental and social concerns. The introductory chapter traces the historical convergence of social work and environmental justice, highlighting key theoretical contributions and culminates in a proposal for a praxis-based eco-social resilience model to guide future interventions. In Chap. 2, the authors introduce the concept of "eco-social life" to challenge anthropocentric paradigms. In Chap. 3, the authors utilize bibliometric analysis to trace the evolution of disaster management research within social work and provides a valuable overview of the scholarly landscape and its emerging trends. The potential of indigenous knowledge and nature-based solutions (NbS) is vividly illustrated in Chap. 4, and global case studies, where NbS successfully blend indigenous knowledge with eco-social work, are also presented. Chapter 5 outlines the critical areas for social work action at micro, mezzo, and macro levels to address contemporary climate vulnerabilities and argues climate change is a real-world problem that needs urgent attention.

A more localized practice perspective is discussed in Chap. 6, where the authors document a citizen-led flood management initiative in Kerala, a southern state in

India. This case study showcases how grassroots technological interventions can build community resilience and strengthen early warning systems and presents a practical example of eco-social work. Chapter 7 explores the combination of traditional forest resource management practices into contemporary micro-level institutions and argues that the tribal leadership and ecological governance embody essential values of eco-social work. An equally innovative example is provided in Chap. 8 that explores the co-production of localized marine forecasts with fishing communities. Treating risk communication as a right is vital to the eco-social approach. The transformative role of technology in eco-social interventions and the tools like GIS, social media analysis, mobile apps, and satellite imaging are examined in Chap. 9. A rights-based framework for ethical and inclusive technology use in disaster contexts is argued for. Mental health, often overlooked in disaster contexts, is the focus of Chap. 10. Through an eco-social lens, the authors interlink trauma-informed care with environmental restoration. In Chap. 11, eco-social perspectives and systems theory are applied to the domain of waste management for a critique of waste management in the in fine-dining restaurants, and advocates for sustainable consumption and just-hospitality practices. The concluding chapter (Chap. 12) synthesizes the cross-cutting themes to discern the future directions of this area, and identifies ten potential topics to guide future research and intervention in eco-social work and DRR domain.

We extend our deepest gratitude to the contributing authors for their intellectual generosity, to the reviewers for their thoughtful engagement, and to the publishing team for their consistent support. We hope that this volume will serve as a resource for building resilient, just, and sustainable communities, and will contribute to the ongoing dialogue on the future of social work in the Anthropocene.

Thiruvananthapuram, Kerala, India

Joice K. Joseph
A. P. Pradeepkumar

Contents

Editors and Contributors

About the Editors

Joice K. Joseph serves as a faculty member in the Master of Social Work (Disaster Management) Department at the Loyola College of Social Sciences, Kerala, India. He holds a multidisciplinary academic background and professional expertise spanning Environmental Science, Social Work, and Disaster Risk Reduction (DRR). He is an established academic and practitioner with a unique interdisciplinary profile, blending qualifications in Environmental Sciences (PhD), Disaster Management (MSc), and Social Work (MSW). With over a decade of academic and field experience, he advances integrated approaches that promote both environmental and social justice through notable articles that examine the intersections of Environmental Sustainability, Disaster Risk Reduction, and Social Equality. He collaborated with both governmental and nongovernmental organizations on various initiatives aimed at implementing DRR strategies that prioritized social and environmental sustainability. In a prior role, he served as Development Coordinator at the M. S. Swaminathan Research Foundation (MSSRF), India. He has also participated in multiple field-based research initiatives, and in 2016, he was a member of the Indian Arctic Expedition, based at Ny-Ålesund, Svalbard, Norway, an opportunity that enhanced his socio-ecological research exposure at a global level. He is the founder and lead of the pioneering research cluster, the Disaster Risk Reduction and Eco-social Work Research Group (DRR & ESWRG), a collective dedicated to advancing scholarly inquiry and practice in these fields. His work provides a vital framework for embedding ecological principles into the core of social work practice and DRR, aiming to build more resilient and just societies.

A. P. Pradeepkumar is Professor of Geology at the University of Kerala, where he serves as the Director of the School of Earth System Sciences. With an illustrious academic background, he holds a BSc, MSc, and PhD in Geology from the University of Kerala. He previously held the position of Reader in Disaster

Management at the School of Environmental Sciences, Mahatma Gandhi University, India. A veteran researcher and academician, his expertise spans diverse fields, including disaster management, GIS, ore genesis, and petrology. His international experience includes significant contributions to Clean Development Mechanism (CDM) and Joint Implementation (JI) projects for carbon credits as part of European Union initiatives at the IER, Germany. He is also the visionary behind the Disaster, Risk and Vulnerability Conference series, which he initiated in Kerala in 2011, now a reputable forum in its sixth edition. Before transitioning to academia, he served as a hydrogeologist with the Central Ground Water Board of India, bringing practical field experience to his teaching and research endeavors. With over 25 years of experience, he continues to mentor and inspire the next generation of geoscientists.

Contributors

S. Abhilash Department of Atmospheric Sciences, Cochin University of Science and Technology (CUSAT), Kochi, Kerala, India

Advanced Centre for Atmospheric Radar Research (ACARR), CUSAT, Kochi, Kerala, India

T. Adarsh Department of Geography, Sree Sankaracharya University of Sanskrit, Kalady, Ernakulam, Kerala, India

S. C. Andrew Michael Loyola College of Social Sciences, Thiruvananthapuram, Kerala, India

Arun Anil Loyola College of Social Sciences (Autonomous), Thiruvananthapuram, Kerala, India

K. P. Arunjith Department of Humanities and Social Sciences, Indian Institute of Space Science and Technology, Thiruvananthapuram, Kerala, India

Sebin Thomas Babu St. Ignatius of Loyola College, Kaunas, Lithuania

K. R. Baiju School of Environmental Sciences, Mahatma Gandhi University, Kottayam, Kerala, India

Shahnaz Basheer Loyola College of Social Sciences (Autonomous), Thiruvananthapuram, Kerala, India

Vinumol Devassy Department of Sociology, Loyola College of Social Sciences (Autonomous), Thiruvananthapuram, Kerala, India

Kristina Grumadaite St. Ignatius of Loyola College, Kaunas, Lithuania

K. P. Haneena Sulthana Department of Psychology, Sree Sankaracharya University of Sanskrit, Kalady, Ernakulam, Kerala, India

Angelika Iser Fakultat fur angewandte Sozialwissenschaften Hochschule München University of Applied Sciences, Munich, Germany

Andria John Department of Social Work, Central University of Kerala, Kasaragod, Kerala, India

Sonny Jose International Relations and Mobility | Loyola College of Social Sciences, Sreekariyam, Thiruvananthapuram, Kerala, India

Joice K. Joseph Department of MSW Disaster Management, Loyola College of Social Sciences (Autonomous), Thiruvananthapuram, Kerala, India

Shibu K. Mani Department of Civil Engineering, School of Engineering and Technology, CHRIST (Deemed to be University), Bengaluru, Karnataka, India

Max Martin Department of Life Sciences, Christ University, Bengaluru, Karnataka, India

Centre for Landscapes, Wildlife and Marine Ecology (CLIME), Christ University, Bengaluru, Karnataka, India

Angelo P. Mathew Human Resource Management, Loyola College of Social Sciences, Thiruvananthapuram, Kerala, India

Neenu Mathews Inter-University Centre for Disability Studies, Mahatma Gandhi University, Kottayam, Kerala, India

Lekshmi V. Nair Department of Humanities and Social Sciences, Indian Institute of Space Science and Technology, Thiruvananthapuram, Kerala, India

Megha N Nair School of Environmental Sciences, Mahatma Gandhi University, Kottayam, Kerala, India

Irish M. Polinga College of Engineering, Architecture and Technology, De La Salle University, Dasmariñas, Philippines

A. P. Pradeepkumar Department of Geology, University of Kerala, Thiruvananthapuram, Kerala, India

M. Pushpam Department of Sociology, University of Kerala, Thiruvananthapuram, Kerala, India

Laxmi Putran Department of Social Work, Central University of Kerala, Kasaragod, Kerala, India

M. Riswan Department of Sociology, South Eastern University of Sri Lanka, Oluvil, Sri Lanka

S. P. Sreya Department of MSW Disaster Management, Loyola College of Social Sciences, Thiruvananthapuram, Kerala, India

Katja Stoppenbrink Fakultät für angewandte Sozialwissenschaften | Department of Applied Social Sciences, Hochschule München University of Applied Sciences, Munich, Germany

P. Vijayakumar Department of Environmental Sciences, University of Kerala, Thiruvananthapuram, Kerala, India

Chapter 1
Social Work, Environment and Disaster Risk Reduction

Joice K. Joseph and A. P. Pradeepkumar

Abstract Environmental challenges in the contemporary world are posing significant vulnerabilities to society. Disasters are not simply "natural" phenomena; they are truly socially manufactured events in which preexisting social, economic, and political injustice cause vulnerable communities to lose their homes and means of livelihood at disproportionate levels. Ecological crises are one of the worst threats in the modern-day world, from pollution to the climate crisis. This introductory chapter explores the critical nexus of social work and environment in the context of Disaster Risk Reduction (DRR), emphasizing the interrelationship between social justice, environmental justice, and community resilience. This study investigates the key themes and concepts related to the physical and social environment, social work and environment, DRR and social work practice, contributions of Lena Dominelli and Margaret Alston, and the nexus of social and environmental justice and finally introduces a praxis-based eco-social resilience model. In order to effectively address the social and ecological vulnerabilities exacerbated by ongoing environmental concerns, social work and environmental justice must be integrated.

Keywords Environmental social work · Community resilience · Praxis-based eco-social resilience model · Disaster risk reduction

J. K. Joseph (✉)
Department of MSW Disaster Management, Loyola College of Social Sciences (Autonomous), Thiruvananthapuram, Kerala, India

A. P. Pradeepkumar
Department of Geology, University of Kerala, Thiruvananthapuram, Kerala, India

J. K. Joseph, A. P. Pradeepkumar (eds.), *Eco-Social Work for Disaster Risk Reduction*, Disaster Resilience and Green Growth, https://doi.org/10.1007/978-981-95-4885-9_1

1.1 Introduction

The profession of social work, which began as a structured response to the socio-economic changes of the industrial revolution that grew from its foundations in charity and philanthropy, has transformed into an exceptional professional field that has evolved from welfare to empowerment. The development of making social work a known profession started in Western nations in the late 1800s, and during this time, university education, secular values, and evidence-based approaches were familiarized to the academic field (Jennissen and Lundy 2011). While the challenges of earlier periods were shaped by the forces of industrialization and capitalism, the contemporary society is confronted with emerging complexities (McCaffery and Boetto 2024; Alston 2015; Dominelli 2011, 2013). In the context of climate change, social work must move from a purely human-centric paradigm to an eco-social one. This acknowledges that climate change is a crisis multiplier that exacerbates preexisting vulnerabilities and forces practitioners to address the interdependence of ecological and human well-being from a macro, systems-oriented perspective rather than micro-thinking. Figure 1.1 attempts to illustrate a range of socio-environmental concerns characterizing the contemporary world. This introductory chapter examines the emerging interdisciplinary field of eco-social work as a transformative approach to disaster management, highlighting the vital interconnections between human well-being, environmental sustainability, social justice, and the health of natural ecosystems.

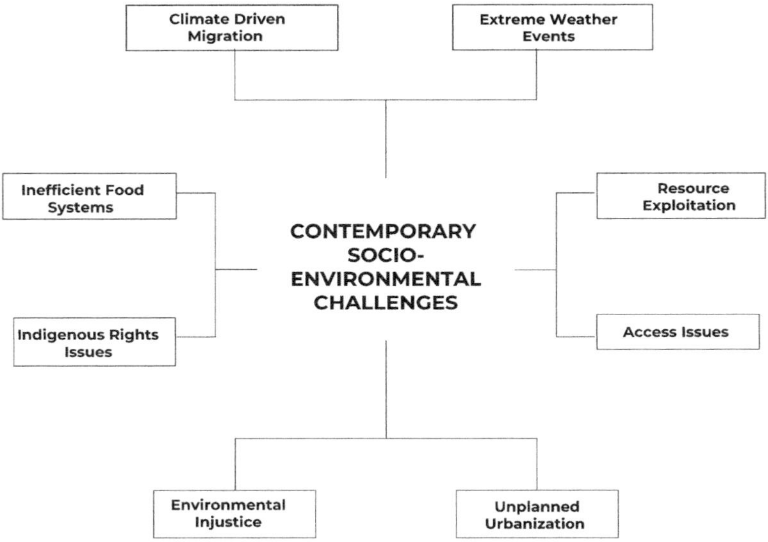

Fig. 1.1 Major contemporary socio-environmental challenges

1.2 The Physical and Social Environment

Human well-being is greatly impacted by both the social and physical environments, and is deeply interconnected. While the social environment consists of the relational, cultural, and socioeconomic contexts that impact norms, support, and opportunities, the physical environment consists of the real, natural, and generated features that influence health behaviours and well-being (Kashima 2020). All of these variables collectively form different but connected determinants of individual and societal wellness.

The physical environment is generally tangible spaces and surroundings. This includes all living (Biosphere) and non-living (Geosphere) things, and within the natural environment, humans construct distinct and varied social environments (Wheeler et al. 2015). Health and quality of life are fundamentally influenced by the physical environment (WHO 2016). It has a foundational role in health and quality of life. Physical and mental health outcomes are influenced by access to green places, clean air, water, and sanitary facilities. Accessible, secure, and appealing surroundings, as well as those that are congested, dirty, and devoid of green space, can have a significant impact on people's health, behaviour, and general well-being, frequently exacerbating disparities across various social groups (Kim and Hyun 2018; Kashima 2020).

Relationships, social networks, protection, and cohesiveness within communities are all parts of the social environment (Kawachi and Berkman 2014). Most of them are intangible. The networks and interactions we have with others, including family, friends, neighbours, and colleagues, as well as cultural and societal variables like common values, beliefs, and traditions, as well as further broader social and economic processes like socioeconomic status, racial relations, and power dynamics, are all considered to be part of the social environment. These factors have a significant impact on individual behaviour, opportunities, and general well-being. Negative social environments characterised by isolation or prejudice can result in stress, depression, and negative behaviours, whereas supportive ones foster resilience and emotional well-being (Bricout and Gray 2006). In the face of hardship, whether during natural disasters or public health emergencies, communities with robust support networks and social cohesion typically exhibit greater resilience.

Physical features can help or impair social relationships in addition to enabling an active lifestyle. Similarly Social settings can amplify the impact of physical environments too. Human behaviours and social structures alter the physical environment, while the physical environment offers the resources and limitations that mould social interactions (Kemp 2011). This co-creation affects community development, health, and well-being. Problems can be reduced or increased by the interaction of the social and physical settings. Poor urban design and environmental neglect, especially in low-income regions, can result in dangerous living circumstances that affect social outcomes like employment and educational achievement as well as physical safety, as Evans (2003) highlighted. According to research, social and physical settings both affect health, with the positive effects of healthy physical surroundings being amplified by supporting social networks (Gray et al. 2013; Dylan 2013; Dominelli 2012a).

1.2.1 *One Health Approach: Collaborative Action for Global Well-Being*

The FAO, UNEP, WHO, and WOAH are the quadripartite agencies that developed the One Health approach, which emphasises the interdependence of environmental, animal, and human health and its significance for global well-being. This emphasises the necessity of strong monitoring systems, policy integration, and community engagement in order to address complex health concerns such as zoonotic diseases, antibiotic resistance, and climate change in a comprehensive, cooperative, and flexible manner (FAO et al. 2022). The One Health approach is based on the principle that the health of humans, animals, and the environment is interconnected, and the well-being of one depends on the health of all. In the context of physical environmental concerns, this is especially clear. For humans to solve the contemporary global socio-environmental challenges, the One Health approach is especially relevant.

1.2.2 *Mission LiFE (Lifestyle for Environment) Initiative*

The Indian government's 2021 LiFE Campaign is an example of a grassroots environmental movement that encourages people to live sustainably and in harmony with the planet's natural systems. The campaign emphasizes the need for social environment changes, including individual and community behavior, in addition to policy-level interventions for environmental degradation and climate change, which are serious physical environment issues (MoEF 2022). The campaign uses social responsibility to enhance ecological results and advance sustainable living by promoting eco-friendly daily behaviors. The campaign's mobilization of citizens as Pro-Planet People fosters a socially engaged atmosphere that normalizes and promotes sustainable practices. By lowering pollution, conserving resources, and encouraging sustainable consumption, it also seeks to restore and safeguard the physical environment.

1.3 Social Work and Environment

The physical environment is highly relevant to all methods of social work—at the micro-, meso-, and macro-levels—ranging from work with individuals to broader social action. But social work is traditionally considered as human-centered—anthropocentric (Boetto 2017). However, addressing contemporary global challenges such as climate change, which leads to the Anthropocene, requires a paradigm shift toward an eco-social approach that comprehensively integrates the physical environment as an essential component of human well-being, recognizing humans as inseparable from and dependent upon the natural world for sustainable justice and resilience. In order to address today's ecological and social issues, social work, which has historically concentrated on human well-being, must broaden its focus to include environmental support as a component of holistic assistance. This is because environmental health is closely linked to the health of individuals and communities.

The International Federation of Social Workers (2012), along with the International Association of Schools of Social Work and the International Council on Social Welfare, called on social workers to collaborate with other professionals to promote environmental sustainability for current and future generations and they also listed environmental sustainability as a key focus area for future activities. In 2015, the Council on Social Work Education (CSWE) in the US included environmental justice as part of one of its core competencies (CSWE 2015). They have added (included) environmental justice to competencies 3 and 5.

More intensive research aimed at developing conceptual frameworks to understand social work's relationship with the environment emerged during the 1990s, with notable contributions from Kemp et al. (1997), the eco-social framework of Coates (2003), Kemp (2011), and Lena Dominelli (2012a), marking a significant shift toward integrating ecological perspectives into social work theory and practice. In 1999, the NASW declared environmental exploitation as a violation of social justice, and in 2010, the CSWE identified sustainability as the defining social justice issue of the new century, emphasizing the integral role of environmental stewardship in social work practice (Dewane 2011).

1.3.1 Environmental Social Work

Mainly focussing on the impact of environmental issues on human well-being (e.g., pollution and human health). It typically involves adapting traditional social work methods to respond to these human-centered environmental challenges. Environmental social work describes the means and approaches through which social workers attempt to resolve or avoid social issues and encourage progress by interacting with the environment through a give-and-take and sustainable processes, which is premised on ecological justice, sustainability, and the pursuit of environmental and social justice as related objectives (Gray et al. 2013; Ramsay and Boddy 2016). Environmental social work cuts across an enormous variety of applications, ranging from environmental education, activism, community work, disaster relief, and policy-making.

1.3.2 Ecological Social Work

The Life Model of Social Work Practice, grounded in ecological systems theory, examines how individuals interact with different layers of their environment, such as family, community, and society. However, it has been criticized for not fully incorporating the biophysical environment and for remaining largely anthropocentric, even though it became a foundational framework for ecological social work (Alston et al. 2016).

1.3.3 Eco-Social Work

Natural environment is seen as a core part of the "person-in-environment" perspective and a key focus of justice. It promotes an ecocentric view, recognizing that humans are deeply interconnected with the entire ecosystem. Eco-social Work places strong emphasis on macro-level interventions and structural change, understanding that environmental problems are closely tied to broader societal, economic, and political systems. It has a broader and global coverage and research agenda (Matthies et al. 2001; Molyneux 2010). Eco-social workers advocate for a critical reconceptualization of social work's traditional engagement with environmental issues, asserting that the profession must move beyond its conventional focus on social contexts to effectively respond to the pervasive challenges posed by contemporary ecological crises. They emphasize the urgent need to integrate ecological and sustainability principles more comprehensively into social work education and pedagogy, thereby reshaping the discipline to address interconnected social and environmental injustices (Besthorn 2012; Coates and Gray 2012; Coates 2005).

Eco-social work practices in India are characterized by collective social action movements that advocate for people's rights over natural resources, largely reflecting the environmentalism of the poor. Notable examples include the Chipko Movement (1973), Appiko Movement (1983), Narmada Bachao Andolan (1985), Silent Valley Movement (1973), Gadhamardan Bachao Andolan (1982), Chilika Bachao Andolan (1990s), and various urban-based environmental movements (Biswal 2023).

Among the emerging trends in global social work, some of the most significant green approaches are sustainable social work (Mary 2008), eco-social practice (Peeters 2011), human self and eco-self (Besthorn 1997), ecological feminism—ecofeminism (Griffin 1990), green social work (Dominelli 2012a), deep-ecological social work (Besthorn 2000), ecological literacy (Fleischer 2011), ecological suicide or ecocide (Diamond 2005), eco-emotions (McCaffery and Boetto 2024), and eco-spiritual social work (Besthorn et al. 2010), among others, each proposing special frameworks for the integration of ecological views into social work theory, practice, and advocacy.

1.4 Major Theories and Models of Social Work and Environment

Many theories, such as systems theory, often focus on human perspectives and don't adequately address the physical environment. While environmental social work is influenced by models like PIE and ecological systems theory, newer approaches such as the Transformative Eco-Social Model promote an ecocentric view by combining sustainability principles and deep ecology.

1.4.1 Person in Environment Theory (PIE)

The roots of the Person-in-Environment (PIE) theory were mainly influenced by Mary Richmond, whose book Social Diagnosis (Richmond 1917) laid the groundwork for this approach. According to Mary Richmond, social workers should consider a person's relationships, environment, and available resources in addition to their personal issues if they want to provide them with genuine assistance. She emphasized that understanding how a person interacts with their environment is key to finding effective solutions and offering meaningful support. For many decades, the person-in-environment model has been widely accepted as a central framework in social work theory and practice, often without much critical examination and unquestioned (Germain 1979). Eco-social workers use an ecocentric lens to expand upon the person-in-environment theory, integrating the well-being of the natural world with that of human beings in their practice. PIE mostly considers diverse domain areas of social, political, familial, temporal, spiritual, economic, and physical environments of a person (Kondrat 2011). But the PIE theory often neglects the environment-in-person (Dewane 2011).

In the 1990s, James Karls and Karen Wandrei introduced the PIE Classification System in collaboration with the National Association of Social Workers (NASW) to provide a structured approach for assessing clients beyond traditional medical or psychiatric models (Karls et al. 1994). This four-factor PIE classification offers a holistic framework for understanding problems, guiding assessments, and planning interventions. This classification clearly addresses the aspects of the client's physical and social environment—like the quality of their housing, neighborhood safety, availability of resources, and potential exposure to hazards—that could affect their ability to function and thrive.

Kemp et al. (1997) introduced the concept of environmental intervention, describing it as both direct action within the environment and the transformative process of critically analyzing how environmental conditions shape individual and collective perspectives. This concept is closely aligned with person-in-environment theory, which emphasizes the dynamic interplay between an individual's concerns, resources, and capabilities, and the shared actions undertaken within their community to address environmental challenges. By integrating these perspectives, practitioners can more effectively support individuals and groups in understanding and responding to the complex realities of their environments.

1.4.2 Deep Ecology

The term "deep ecology" was introduced by a Norwegian philosophy professor, Arne Næss, in 1973 (Mathews 2001; Næss 1973). It advocates for a shift from human-centered (anthropocentric) views to an ecocentric perspective. Nature is not a resource for human exploitation; all organisms have intrinsic value and worth of life. Næss cites Mahatma Gandhi, the father of India's independence, as a key inspiration for his ecological thought, highlighting that Gandhi's final goal was not merely political freedom for India, but the deeper liberation of the individual, exceeding the Western notion of the self as a separate, isolated ego to a universal self (Naess 1995). Gandhi's Sarvodaya philosophy emphasizes the well-being of all humans and other living beings, urging people to maintain a close and caring relationship with the non-human world (Gandhi 1961). Deep ecologists argue that conventional human rights-based social justice is insufficient, as it narrowly focuses on human-to-human recognition and distribution. Instead, they advocate for a deeper form of justice that intrinsically links social and ecological well-being, recognizing their profound interconnections in the twenty-first century (Besthorn 2002, 2003; Drengson 2005).

1.4.3 Ecological Systems Theory

Bronfenbrenner's Ecological Systems Theory (1977) explains that human development is shaped by interactions within five interconnected environmental systems: the microsystem, mesosystem, exosystem, macrosystem, and chronosystem. While the theory acknowledges both social and physical influences, it originally placed less emphasis on the physical environment. Later scholars, such as Sheldrick et al. (2019), have expanded the theory to better address the role of the physical environment, especially in areas like social work, environmental justice, and disaster resilience.

1.4.4 The Life Model of Social Work Practice

Carel Germain and Alex Gitterman are the pioneers of the Life Model of Social Work Practice, which emphasizes the interaction between people and their environments across the life course (Gitterman and Germanin 1976). Life model is fundamentally connected to the physical environment through its ecological perspective. The developers of the life model emphasises that a person's well-being is shaped not only by their relationships and social systems but also by their physical surroundings (natural and built environment).

1.4.5 The Transformative Eco-Social Work Model

Heather Boetto, of Charles Sturt University, argues for a paradigmatic shift toward an ecologically centered approach and proposed a transformative eco-social model that is congruent across ontological (being), epistemological (thinking), and methodological (doing) dimensions of practice (Boetto 2017). Boetto suggests that eco-social work should operates on three levels: the personal or "self" level (which includes eco-friendly lifestyles, consumption practices, and personal connections with nature), the theoretical or "thinking" level (which involves reflecting on eco-system justice, traditional knowledge, and socio-political systems), and the practice or "doing" level (which focuses on community interventions and social actions for eco-social change). Together, these levels guide social work practitioners in integrating ecological and social justice principles into their work. The transformative eco-social work model is one of the latest and most comprehensive approaches, offering direct practical applications.

Alongside well-established frameworks in environmental social work, radical environmentalism stands out as a key paradigm that is distinguished by its use of direct-action strategies to address ecological degradation. It is ethically and philosophically based in deep ecology and ecocentrism, which together offer a strong basis for opposing dominant anthropocentric paradigms and promoting significant systemic change. Radical environmental and grassroots movements include the Earth First! Movement, which was established in the southwestern United States in 1980 and pioneered direct action tactics like tree-sitting to protect wilderness areas; militant offshoots like the Animal Liberation Front (ALF) and Earth Liberation Front (ELF); contemporary climate activism exemplified by mass civil disobedience campaigns like Just Stop Oil; and historically significant indigenous and local movements like the Chipko and Appiko Movements in the Indian Himalayas and Western Ghats in the 1970s and 1980s, as well as the Bishnoi Movement in Rajasthan in the eighteenth century. Eco-activism, Eco-philosophy, Ecological consciousness, grassroot environmentalism, etc., are also other key terminologies in this direction.

1.5 Anthropocentric to Ecocentric

Social work's growing awareness of the connection between human and environmental well-being is demonstrated by the combination of environmental and social work theories and models, from Person-in-Environment (PIE) to Transformative

Eco-Social Work Model and More established frameworks like ecological systems theory lay the groundwork by emphasising the social context of individuals, while more recent models expand beyond anthropocentric perspectives to include ecocentric ethics, environmental justice, and sustainability. By supporting radical environmentalism, deep ecology, and the value of all living forms, the most recent iterations oppose the predominance of human-centered paradigms and encourage systemic change through direct action and grassroots activism. Social workers have the ability to advance justice and holistic well-being, which benefits not only individuals and communities but also the ecology of the entire planet.

1.6 Disaster Risk Reduction and Social Work Practice

Disasters are increasing globally but their effects differ with vulnerabilities. The uneven distribution of disaster effects arises from entrenched social, economic, and political structures, whereby preexisting inequalities among rich and poor populations condition both vulnerability and recovery, revealing that disasters are fundamentally socio-political events shaped by historical contexts rather than purely natural occurrences (Cutter 2018; Hyndman 2008; Dominelli 2015, 2018). Among other line disciplines, social workers can contribute to the entire disaster management spectrum since they frequently encounter the dimensions of vulnerability as part of their job tasks (Dominelli 2015).

Literature about the involvement of social workers during disasters is not extensive and most of them focus on post-disaster practical service provision (Cooper and Briggs 2014). Studies reported the involvement of settlement workers during the Chicago fire in 1871 and Charity Organization Societies (COS) during the San Francisco Earthquake of 1906 in service coordination and community mobilization, which provided the roots for disaster social work practice (Zakour 1996). There is a need for a human face in disaster management; currently it is more technical. Social workers can bring immense personal and professional resources to the disaster management spectrum, and they can perform crucial roles in the Disaster Risk Reduction (DRR) cycle at all levels (Fig. 1.2).

The first author (JKJ)'s journey toward eco-social work and disaster risk reduction emerged from an interdisciplinary basis, rooted in environmental science and disaster management, acquired through a pioneering Master of Science program at the Mahatma Gandhi University campus in Kerala, India. While engaged deeply in both academic research and field practice within disaster management, I observed firsthand the disproportionate vulnerabilities faced by marginalized communities during any hazards. This practical exposure highlighted the indispensable role of Social Work in addressing the social dimensions of disaster impacts, catalyzing a shift to pursue advanced Social Work education. As a student spanning the fields of Disaster Risk Reduction, environmental studies, and social work, the author

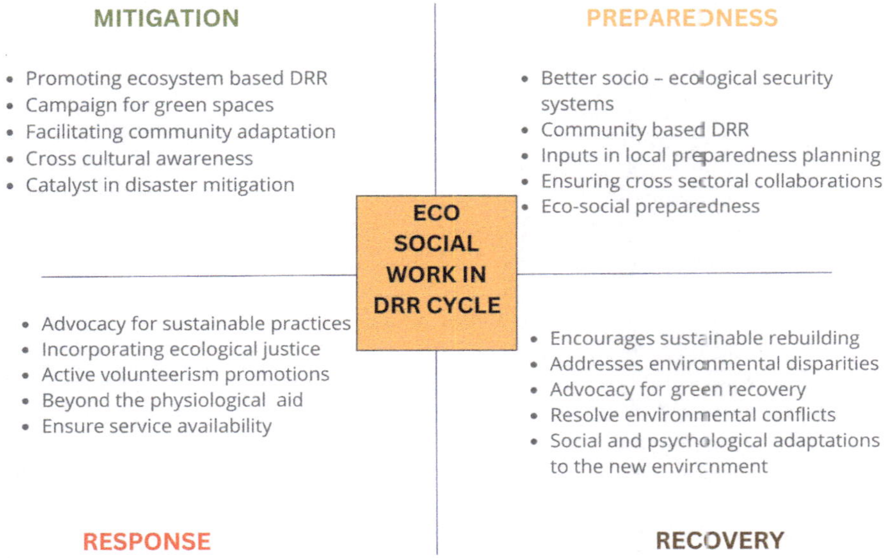

Fig. 1.2 Eco-social work in DRR cycle

recognized that effective DRR necessitates an integrated approach—one that seamlessly links ecological sustainability with social justice and community empowerment. This realization shaped the author's academic and professional trajectory, emphasizing the critical fusion of environmental and social paradigms to enhance resilience and support vulnerable populations throughout disaster contexts.

1.6.1 Iconic Contributions of Lena Dominelli and Margaret Alston to Eco-Social Work and Disaster Risk Reduction

Professors Lena Dominelli and Margaret Alston have made significant advancements in integrating environmental social work to DRR, as highlighted in the subsection that follows. Because of their leadership, eco-social work has emerged as a vital framework for addressing disasters, climate change, and community resilience.

Professor Lena Dominelli, chair in Social Work in the Faculty of Social Sciences at Stirling University, is a pioneering scholar in the emerging field of green social work, community resilience, and environmental justice. She has

(continued)

also served as President of the International Association of Schools of Social Work (IASSW) from 1996 to 2004 and chairs the IASSW Committee on Disaster Interventions, Climate Change, and Sustainability. Dominelli became its first chair and later also chaired the Rebuilding People's Lives After Disasters Network (RIPL), which the IASSW established after the 2004 tsunami. She subsequently served as head of the Disaster Intervention, Climate Change, and Sustainability Committee. She has represented the social work profession at the United Nations discussions on climate change since Cancun, Mexico, in 2010. She has created green social work as a new paradigm for theory and practice. She has published widely in green social work, social work and disaster risk reduction. *Green Social Work from Environmental Crises to Environmental Justice*, published in 2013, is the most notable book of Lena Dominelli. The concept of "green social work" is proposed in this academic work as a holistic approach that blends ecological preservation with human well-being in order to address the interdependencies between people and their sociocultural, economic, and physical surroundings. It criticises industrialisation, consumerism, and the government's incapacity to govern for all, highlighting the ways in which these issues lead to environmental harm, mass migration, social unrest, and the disproportionate suffering of marginalised people. She illustrates how social work may advance social and environmental justice locally, nationally, and globally by highlighting the importance of community-led projects, strong resilience, and equitable resource distribution using examples from catastrophe responses and indigenous perspectives. Professor Lena Dominelli has several publications in this line and is a notable member of the global academic community (Dominelli 2011, 2012a, b, 2013, 2014, 2015, 2018).

Professor Margaret Alston is a renowned academic in the discipline of disaster risk reduction (DRR) within a social work framework, with specific expertise in gender, climate change, and environmental justice. She is a Professor of Social Work and Head of Department at Monash University, Australia. Her extensive body of research has focused on the intersection of climate vulnerability and gender, and she has directed and participated in many international projects, including some in Bangladesh for Oxfam and a seminal 2008 study on gender and climate change in India, written for the Gender Division of the United Nations Food and Agriculture Organization (FAO). Among her notable scholarly works is the co-authored book *Social Work and Disasters: A Handbook for Practice* (2019). The book promotes a comprehensive and multidisciplinary strategy for disaster social work that goes beyond traditional emergency management to include preparedness, response, and sustained recovery with many real-life case studies. It provides an eco-social framework that assists practitioners in incorporating environmental consciousness into disaster interventions. Professor Alston's contribution goes a long way in

(continued)

bridging the important gap in social work literature by highlighting the significance of disaster risk reduction in the context of green social work, focusing particularly on gender dimensions of environmental emergencies. Her works address the complexities of social work disaster practice, especially at macro level interventions in the age of disasters Alston (2013a, b, 2015, 2017; Alston and Besthorn 2012; Alston and Kent 2014; Alston et al. 2016; Alston and Chow 2021).

1.7 Environmental Justice: A Core Dimension of Social Justice

Environmental justice is social justice. Environmental justice is a form of social justice. Environmental hazards do not affect all populations equally; their impacts are shaped by social, physical, and economic factors. Many social issues are directly connected to the environmental status and sustainability. Environment is a common terminology in social and environmental justice but is understood in different ways. The social justice and environmental justice are well connected, especially from a macro-ecological practice perspective (Nesmith and Smyth 2015). Jane Addams can be considered a pioneer in social justice through the formation of Hull House in Chicago in 1889. Social work's historical roots in addressing environmental issues can be found in Jane Addams because she also fought to address the environmental issues of poor people who mainly suffered from the industrial pollutions and lived in vulnerable slum locations (Addams 1961). Hull House tackled societal issues along with degraded urban environmental health for sustainable community development (Tadesse and Elsen 2023; Halatcheva et al. 2024).

Both environmental justice and social justice are for addressing a better quality of life in a sustainable world (Beltran et al. 2016). According to the notable author Fred Besthorn, although social justice advocacy is a fundamental aspect of Western social work, ultimately all forms of justice are ecological (Besthorn 2011). Social workers are traditionally a part of the environmental justice movement (Teixeira and Krings 2015). Minimizing environmental harm and ensuring the fair distribution of natural resources to the greatest number of people are goals that align with both social and environmental justice. The integration of social justice and eco-justice is indispensable for the protection of the rights and dignity of humans and non-human beings alike, reflecting an urgent imperative in today's world where overlying social and environmental crises, especially in the changing climate context (McLaughlin 1993).

Environmental concerns are important for social workers, due to the profession's distinctive emphasis on the person-in-environment. Kemp (2011) characterizes the current era as the "third wave" of environmental social work, emphasizing the integration of ecological and social concerns. In this contemporary context, Kemp argues that environmental justice—intertwined with social justice—should be at the forefront of social work practice. Kemp's observation is crucial for the integration of environmental and social justice within social work practice. Dr. Robert Bullard is widely recognized as the "father of environmental justice" for his pioneering work in documenting and raising awareness about the disproportionate impact of environmental hazards on marginalized communities. The power imbalance in human victims of environmental degradation leads to multifractal injustice; poor and marginalized sections are excessively affected by disaster consequences and are a kind of slow violence (Nixon 2011). Social justice cannot exist without environmental and species justice (Gaard 2011).

Social workers in the contemporary world need an environment-in-the-person approach. Social workers can initiate a virtuous cycle across the micro-, meso-, and macro-systems by building community capacity to resist environmental exploitation, which can lead to effective Disaster Risk Reduction.

1.8 Praxis-Based Eco-Social Resilience—A New Approach

Praxis in any profession is attempting to develop theoretical knowledge into practice (Ritchie 2023). An eco-social resilience framework that integrates Social, Ecological, and Disaster Risk Reduction systems by addressing the underlying causes of vulnerability and promoting disaster risk reduction is necessary to address the mutual vulnerabilities of humans and the environment. Additionally, an eco-social work lens focused on social and environmental justice outcomes is needed to promote resilient communities. The following section aims to address conceptualising praxis-based eco-social resilience (Fig. 1.3).

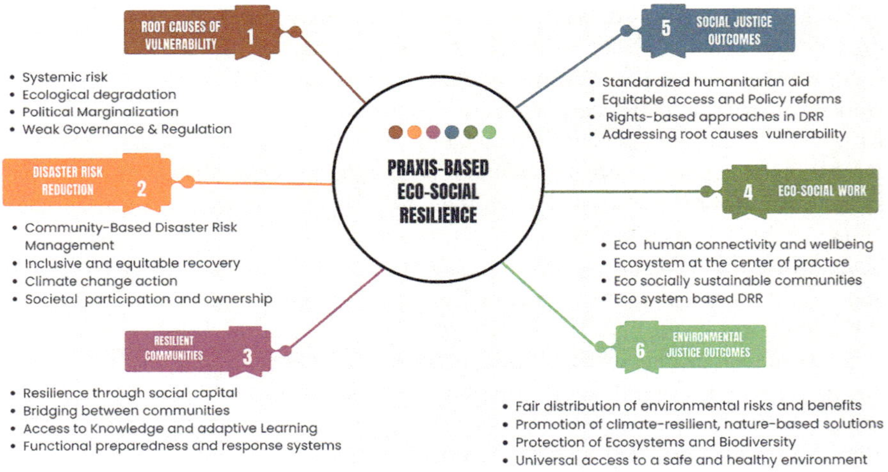

Fig. 1.3 Praxis-based eco-social resilience

The interconnected systemic risk factors are the main root causes of human vulnerability and one of the biggest existential dangers to global society by magnifying cascading disasters (Undheim 2024). Together, ecological degradation, political marginalization and poor governance are the main causes of vulnerability in many communities, which severely compromise societal resilience and adaptability. The increasing number of hazards, from earthquakes to pollution, ultimately end up as ecological catastrophes to nature and the local community (Garnaut 2008). Eco-social work practice provides multi-level approaches (individual, family, community, institutional, structural) to address climate and environmental impacts, including DRR. Community-Based Disaster Risk Management and climate action initiatives are intrinsically aligned with ecocentrism, as they emphasize Ecosystem-Based Disaster Risk Reduction approaches that recognize the interdependence of human resilience and ecological sustainability.

Social justice outcomes, like standardization of humanitarian assistance, equitable access to resources, and rights-based measures, along with environmental justice outcomes including protection of biodiversity and enforcement of the polluter-pays principle, are fundamentally connected with the traits of resilient communities characterized by high social capital, increased cohesion, and fair access to a healthy environment; hence, governments and policymakers should prioritize the eco-social resilience nexus as a determinant framework for practical application, allowing social-ecological systems to better absorb, adapt to, and recover from multidimensional shocks.

1.9 Conclusion

The modern social work profession is increasingly confronted with previously uncommon challenges like the climate crisis, which demand innovative, flexible solutions to effectively navigate the complex and evolving needs of contemporary society. The eco-social resilience nexus presents a transformative paradigm for Disaster Risk Reduction, linking social and environmental justice outcomes to the fundamental characteristics of resilient communities. Climate change and other new global environmental issues require social workers and other academics to think innovatively. The eco-social resilience approach not only mitigates disaster risks but also addresses systemic vulnerabilities by aligning policy and practice with the interdependent nature of social and ecological systems.

References

Addams J (1961) Twenty years at Hull-house. New American Library, a Division of Penguin Group (USA), Inc, New York

Alston M (2013a) Social work in the context of climate change and disasters. In: Connolly M, Harms L (eds) Social work: contexts and practice, 3rd edn. Oxford University Press, South Melbourne, pp 315–326. ISBN 0195520181

Alston M (2013b) Environmental social work: accounting for gender in climate disasters. Aust Soc Work 66(2):218–233. https://doi.org/10.1080/0312407X.2012.738366
Alston M (2015) Social work, climate change & global co-operation. Int Soc Work 58(3):335–363. https://doi.org/10.1177/0020872814556824
Alston M (2017) Eco social work: reflections from the global south. In: Matthies A-L, Närhi K (eds) The ecological transition of societies: the contribution of social work and social policy. Routledge, Abingdon, pp 91–104
Alston M, Besthorn F (2012) Environment and sustainability. In: Lyons K, Hokenstad T, Pawar M, Huegler N, Hall N (eds) The SAGE handbook of international social work. SAGE Publications, London, pp 56–69. ISBN: 9780857023339
Alston M, Chow AYM (2021) Introduction—when social work meets disaster: challenges and opportunities. Br J Soc Work 51(5):1525–1530. https://doi.org/10.1093/bjsw/bcab154
Alston M, Kent J (2014) Coping with a crisis: human services in times of drought. Rural Soc 14(3):214–227. https://doi.org/10.5172/rsj.351.14.3.214
Alston M, Hazeleger T, Hargreaves D (2016) Social work in post-disaster sites. In: McKinnon J, Alston M (eds) Ecological social work: towards sustainability. Palgrave Macmillan, London, pp 158–174. https://doi.org/10.1007/978-1-137-40136-6_10
Alston M, Hazeleger T, Hargraeaves D (2019) Social work and disasters: a handbook for practice. Routledge, Abingdon
Beltran R, Hacker A, Begun S (2016) Environmental justice is a social justice issue: incorporating environmental justice into social work practice curricula. J Soc Work Educ 52:493–502
Berkes F, Colding J, Folke C (2003) Navigating social-ecological systems: building resilience for complexity and change. Cambridge University Press, Cambridge
Besthorn FH (1997) Reconceptualizing social work's person-in-environment perspective:exploration in radical environmental thought. Unpublished doctoral dissertation. University of Kansas, Lawrence
Besthorn F (2000) Toward a deep-ecological social work: its environmental, spiritual and political dimensions. Spiritual Soc Work Forum 7(2):2–7
Besthorn FH (2002) Radical environmentalism and the ecological self: rethinking the concept of self-identity for social work practice. J Progress Hum Serv 13(1):53–72
Besthorn FH (2003) Radical ecologisms: insights for educating social workers in ecological activism and social justice. Crit Soc Work 3(1):66–106
Besthorn FH (2012) Deep ecology's contributions to social work: a ten-year retrospective. Int J Soc Welf 21(3):248–259
Besthorn FH, Wulff D, George S (2010) Eco-spiritual helping and postmodern therapy: a deeper ecological framework. Ecopsychology 2(1):23–32
Bigonnesse C, Mahmood A, Chaudhury H, Mortenson WB, Miller WC, Martin Ginis KA (2018) The role of neighborhood physical environment on mobility and social participation among people using mobility assistive technology. Disabil Soc 33(6):866–893. https://doi.org/10.1080/09687599.2018.1453783
Biswal M (2023) Environmental social work: a path towards sustainable development. Soc Cult Dev India 3(1):61–76
Bobo K, Kendall J, Max S (2001) Organizing for social change: Midwest academy manual for activists. Seven Locks Press, Santa Ana
Boetto H (2017) A transformative eco-social model: challenging modernist assumptions in social work. Br J Soc Work 47:48–67. https://doi.org/10.1093/bjsw/bcw149
Boetto H (2019) Advancing transformative eco-social change: shifting from modernist to holistic foundations. Aust Soc Work 72(2):139–152. https://doi.org/10.1080/0312407X.2018.1484501
Bricout J, Gray D (2006) Community receptivity: the ecology of disabled persons' participation in the physical, political and social environments. Scand J Disabil Res 8:1–21. https://doi.org/10.1080/15017410500335229
Bronfenbrenner U (1977) Toward an experimental ecology of human development. Am Psychol 32:513–531

Coates J (2003) Ecology and social work: toward a new paradigm. Fernwood Publishing, Blackpoint. ISBN-10: 1552661075
Coates J (2005) The environmental crisis: implications for social work. J Progress Hum Serv 16(1):25–49. https://doi.org/10.1300/J059v16n01_03
Coates J, Gray M (2012) The environment and social work: an overview and introduction. Int J Soc Welf 21(3):230–238
Cooper L and Briggs L (2014) 'Do we need specific disaster management education in social work?,The Australian Journal of Emergency Management 29: 38–42.
Cote M, Nightingale AJ (2012) Resilience thinking meets social theory: situating social change in social-ecological systems (SES) research. Prog Hum Geogr 36(4):475–489
Council on Social Work Education (2015) Educational policy and accreditation standards. Retrieved from https://www.cswe.org/getattachment/Accreditation/Standards-and-Policies/2015-EPAS/2015EPASandGlossary.pdf.aspx. ISBN 978-0-87293-209-8
Cutter SL (2018) Compound, cascading, or complex disasters: what's in a name? Environ Sci Policy Sustain Develop 60:16–25. https://doi.org/10.1080/00139157.2018.1517518
Demaine J (2005) Education and global citizenship. Int J Divers Organ Community Nations 5(3):103–109
Dewane CJ (2011) Environmentalism & social work: the ultimate social justice issue, social work today. https://www.socialworktoday.com/archive/092011p20.shtml. Accessed 9 July 2025
Diamond J (2005) Collapse: how societies choose to fail or succeed. Viking Press, New York
Dominelli L (2011) Climate change: social workers contributions to policy and practice debates. Int Soc Work 20(4):430–439. https://doi.org/10.1111/j.1468-2397.2011.00795.x
Dominelli L (2012a) Green social work: from environmental crises to environmental justice. Polity Press, London. ISBN: 978–0–745-65400-3
Dominelli L (2012b) Mind the gap: built infrastructures, sustainable caring relations and resilient communities in extreme weather events. Aust Soc Work 66(2):204–217. https://doi.org/10.1080/0312407X.2012.708764
Dominelli L (2013) Environmental justice at the heart of social work practice: greening the profession. Int J Soc Welf 22(4):431–439. https://doi.org/10.1111/ijsw.12024
Dominelli L (2014) Promoting environmental justice through green social work practice: a key challenge for practitioners and educators. Int Soc Work 57(4):338–345. https://doi.org/10.1177/0020872814524968
Dominelli L (2015) The opportunities and challenges of social work interventions in disaster situations. Int Soc Work 58(5):659–672. https://doi.org/10.1177/0020872815598353
Dominelli L (2018) Green social work and the uptake by the Nepal School of Social Work: building resilience in disaster-stricken communities. In: Evolving narratives of hazard and risk. Palgrave Macmillan, Cham. https://doi.org/10.1007/978-3-319-65211-5_9
Drengson A (2005) The life and work of Arne Naess: an appreciative overview. The Trumpeter 21(1):5–60. ISSN 0832-6193
Dylan A (2013) Environmental sustainability, sustainable development and social work. In: Gray M, Coates J, Hetherington T (eds) Environmental social work. Routledge, New York, pp 62–87
Evans GW (2003) The built environment and mental health. J Urban Health 80(4):536–555. https://doi.org/10.1093/jurban/jtg063
FAO, UNEP, WHO and WOAH (2022) One Health Joint Plan of Action (2022–2026). Working together for the health of humans, animals, plants and the environment. WHO, Rome
Fleischer S (2011) Emerging beliefs frustrate ecological literacy and meaning making for students. Cult Stud Sci Educ 6:235–241
Gaard G (2011) Ecofeminism revisited: rejecting essentialism and re-placing species in a material feminist environmentalism. Fem Form 23:26–53
Gandhi MK (1961) In: Kher VB (ed) In search of the supreme, compiled and, vol 1. Navajivan Publishing House, Ahmedabad
Garnaut R (2008) The Garnaut climate change review: final report. Cambridge University Press, Cambridge, UK

Germain C (1979) Introduction: ecology and social work. In: Germain C (ed) Social work practice, people and environments: an ecological perspective. Columbia University Press, New York, pp 3–22

Germain CB (1981) The physical environment and social work practice. In: Maluccio AN (ed) Promoting competence in clients. Free Press, New York, pp 103–124

Gitterman A, Germain C (1976) Social work practice: a life model. Soc Serv Rev 50:601–610. https://doi.org/10.1086/643430

Gray M, Coates J, Hetherington T (eds) (2013) Environmental social work. Routledge, London/ New York

Griffin S (1990) Curves along the road. In: Reweaving the world: the emergence of ecofeminism. Sierra Club Books, pp 87–99

Haines A, Kovats RS, Campbell-Lendrum D, Corvalan C (2006) Climate change and human health: impacts, vulnerability, and mitigation. Lancet 367:2101–2109

Halatcheva T, Poferl M, Angelika (2024) Jane Addams and the Hull-House Maps & Papers. A knowledge-culture approach. Kieler sozialwissenschaftliche Revue Internationales Tönnies-Forum 2:7–18. https://doi.org/10.3224/ksr.v2i1.02

Healy LM (2008) International social work: professional action in an interdependent world, 2nd edn. Oxford University Press, New York. https://doi.org/10.4060/cc2289en

Hyndman J (2008) Feminism, conflict and disasters in post-tsunami Sri Lanka. Gend Technol Dev 12(1):101–121

IASC (2014) Review of the implementation of the IASC guidelines on mental health

International Federation of Social Workers, International Association of Schools of Social Work, & International Council on Social Welfare (2012) The global agenda for social work and social development: commitment to action. Retrieved from http://cdn.ifsw.org/assets/globalagenda2012.pdf

JennissenT and Lund C (2011) One Hundred Years of Social Work-A History of the Profession in English Canada, 1900-2000. Wilfrid Laurier University Press, ISBN 9781554582808

Karls JM, Wandrei KE (eds) (1994) Person-in-Environment (PIE) system: the PIE classification system for social functioning problems. NASW Press, Washington, DC. ISBN 0-87101-254-5

Kashima Y (2020) Ecologizing social psychology: the physical environment as a necessary constituent of social processes. Personal Soc Psychol Rev 24(2):178–203. https://doi.org/10.1177/1088868319845938

Kawachi I, Berkman LF (2014) Social capital, social cohesion, and health. In: Berkman LF, Kawachi I, Glymour MM (eds) Social epidemiology, 2nd edn. Oxford University Press. https://doi.org/10.1093/med/9780195377903.003.0008

Kemp SP (2010) Place matters: toward a rejuvenated theory of environment for social work practice. In: Borden W (ed) Reshaping theory in contemporary social work: toward a critical pluralism in clinical practice. University of Chicago Press, Chicago, pp 114–145

Kemp SP (2011) Recentring environment in social work practice: necessity, opportunity, challenge. Br J Soc Work 41:1198–1210

Kemp SP, Whittaker JK, Tracy EM (1997) Person-environment practice: the social ecology of interpersonal helping. Aldine de Gruyter, New York

Kim B, Hyun HS (2018) Associations between social and physical environments, and physical activity in adults from urban and rural regions. Osong Public Health Res Perspect 9(1):16–24. https://doi.org/10.24171/j.phrp.2018.9.1.04

Kondrat ME (2011) Person-in-environment. In: Oxford bibliographies. Oxford University Press, Oxford

Mary LN (2008) Social work in a sustainable world. Lyceum Books, Chicago

Mathews F (2001) Deep ecology. In: Jamieson D (ed) A companion to environmental philosophy. Blackwell, pp 218–232. https://doi.org/10.1002/9780470751664.ch15

Matthies A-L, Närhi K, Ward D (eds) (2001) The eco-social approach in social work. University of Jyväskylä. ISBN 951-39-1834-3

McCaffery J, Boetto H (2024) Eco-emotional responses to climate change: a scoping review of social work literature. Br J Soc Work 55:120–140. https://doi.org/10.1093/bjsw/bcae129
McKinnon J, Alston M (eds) (2016) Ecological social work: towards sustainability. Palgrave Macmillan, London. https://doi.org/10.1177/0261018316668067
McLaughlin A (1993) Regarding nature: industrialism and deep ecology. State University of New York Press. ISBN-10-9780791413845
MoEF – Ministry of Environment, Forest and Climate Change, Government of India (2022) Mission LiFE – Lifestyle for Environment [Brochure]. https://static.pib.gov.in/WriteReadData/specificdocs/documents/2022/nov/doc2022119122601.pdf
Molyneux R (2010) The practical realities of eco-social work: a review of the literature. Critical Soc Work 11:61–69. https://doi.org/10.22329/csw.v11i2.5824
Munro S (2012) Rich land, wasteland: how coal is killing Australia. Macmillan, Sydney
Naess A (1995) Self-realization: an ecological approach to being in the world. In: Sessions G (ed) Deep ecology for the 21st century: readings on the philosophy and practice of the new environmentalism. Shambhala, Boston, pp 225–239
Nasess, A. (2017). The shallow and deep, long - range ecology movement. A summary. In the ethics of the environment (pp. 115–120). Routledge
Nesmith A, Smyth N (2015) Environmental justice and social work education: social workers' professional perspectives. Soc Work Educ 35:484–501. https://doi.org/10.1080/02615479.2015.1063600
Nixon R (2011) Slow violence and the environmentalism of the poor. Harvard University Press, Cambridge, MA
Noble C (2016) Green social work-the next frontier for action. Soc Altern 35(4):14–19
Peeters J (2011) The place of social work in sustainable development: towards eco-social practice. Int J Soc Welf 21(3):287–298. https://doi.org/10.1111/j.1468-2397.2011.00856.x
Philip D, Reisch M (2015) Rethinking social work's interpretation of 'environmental justice': from local to global. Soc Work Educ 34(5):471–483. https://doi.org/10.1080/02615479.2015.1063602
Ramsay S, Boddy J (2016) Environmental social work: a concept analysis. Br J Soc Work 47:Bcw 078. https://doi.org/10.1093/bjsw/bcw078
Richmond ME (1917) Social diagnosis. Russell Sage Foundation, New York
Ritchie L (2023) Practice of praxis. Journal od adult learning, Knowldge and Inoovation. 6.10.1556/2059.2023.00066
Sheldrick MP, Maitland C, Mackintosh KA, Rosenberg M, Griffiths LJ, Fry R, Stratton G (2019) Associations between the home physical environment and children's home-based physical activity and sitting. Int J Environ Res Public Health 16(21):4178
Tadesse ME, Elsen S (2023) The social solidarity economy and the hull-house tradition of social work: keys for unlocking the potential of social work for sustainable social development. Soc Sci 12:189. https://doi.org/10.3390/socsci12030189
Teixeira S, Krings A (2015) Sustainable social work: an environmental justice framework for social work education. Soc Work Educ 34:513–527
Undheim AT (2024) An interdiciplinary review of systemic risk factors leading up to existential risks, Prog in Dis Sci 22, 2024. https://doi.org/10.1016.j.pdisas.2024.1000326
Wheeler BW, Lovell R, Higgins SL, White MP, Alcock I, Osborne NJ, Husk K, Sabel CE, Depledge MH (2015) Beyond greenspace: an ecological study of population general health and indicators of natural environment type and quality. Int J Health Geogr 14:17. https://doi.org/10.1186/s12942-015-0009-5
World Health Organization (2016) Urban green spaces and health: a review of evidence. WHO Regional Office for Europe. Available at https://www.who.int/europe/publications/i/item/WHO-EURO-2016-3352-43111-60341
Zakour MJ (1996) Disaster research in social work. J Soc Serv Res 22(1–2):7–25

Chapter 2
Eco-Social Life: The Symbiosis Between Society and Nature

Sonny Jose, Katja Stoppenbrink, S. C. Andrew Michael, and Angelika Iser

Abstract We live in an era marked by ecological crises, social fragmentation, and intensifying climate-related disasters, making it imperative to understand the reciprocal relationships between the social and natural worlds. The present chapter explores the concept of eco-social life, a framework that acknowledges the interconnectedness of human societies and ecosystems, and emphasizes mutual dependence, resilience, and sustainability. Moving beyond anthropocentric paradigms, the present chapter attempts an exploration into how social practices, institutions, and values shape, and in turn are shaped by the ecological systems they inhabit.

Keywords Eco-social life · Ecological crisis · Climate change · Ecosystems · Symbiosis

---

S. Jose
International Relations and Mobility | Loyola College of Social Sciences, Sreekariyam, Thiruvananthapuram, Kerala, India
e-mail: sonny.jose@loyolacollegekerala.edu.in

K. Stoppenbrink
Fakultät für angewandte Sozialwissenschaften | Department of Applied Social Sciences, Hochschule München University of Applied Sciences, Munich, Germany
e-mail: katja.stoppenbrink@hm.edu

S. C. Andrew Michael (✉)
Loyola College of Social Sciences, Thiruvananthapuram, Kerala, India
e-mail: andrewmichael@loyolacollegekerala.edu.in

A. Iser
Fakultat fur angewandte Sozialwissenschaften Hochschule München University of Applied Sciences, Munich, Germany
e-mail: angelika.iser@hm.edu

J. K. Joseph, A. P. Pradeepkumar (eds.), *Eco-Social Work for Disaster Risk Reduction*, Disaster Resilience and Green Growth, https://doi.org/10.1007/978-981-95-4885-9_2

2.1 Introduction

> All that man needs for health and healing, has been provided by God in nature; the challenge of Science is to find it. —Paracelsus

Mankind is living in an era marked by ecological crises, social fragmentation, and intensifying climate-related disasters, making it imperative that we understand the reciprocal relationships between the social and natural worlds. The relationship between mankind and nature is as old as the evolution of life. Human beings live in the lap of nature and constantly interact with it. The natural environment influences the life of human beings in every way—the air they breathe, the water they drink, the food they eat, besides the flow of energy and information. Any drastic change in the natural environment not only results in devastating effects but can also pose a threat to the human race (Hannigan 2006). At the same time, human beings modify their environment as a result of their growth, dispersal, activities, death and decay, etc. Thus, all living beings, including man and their environment, are mutually reactive, affecting each other in a number of ways. The human beings (society) and environment are interrelated and interdependent (Kelemen 2023). It calls for a dynamic equilibrium between the two.

This book chapter explores the concept of eco-social life, a framework that acknowledges the interconnectedness of human societies and ecosystems, and the need to emphasize inter-dependence, resilience, and sustainability. Moving beyond anthropocentric paradigms, this article attempts an exploration into how social practices, institutions, and values shape, and in turn are shaped by the ecological systems they inhabit.

Bridging the insights from systems theory, ecocritical approaches, and person-in-environment (PIE) theory, the present chapter aims to provide a multidisciplinary narrative that is both academically grounded and accessible to practitioners. Drawing on perspectives from both the Global North (e.g., European Union policies and climate resilience strategies) and the Global South (particularly India and Southeast Asia), the chapter highlights culturally and historically embedded relationships with nature, while also addressing contemporary eco-social risks and inequalities.

The chapter further incorporates the *Transformative Eco-social Model*, which advocates for structural change and sustainable social development, integrating ecological and social justice approaches. Here, the authors also engage the strengths-based approach (SBA) to disaster risk reduction (DRR), exploring how "community resilience as well as local and indigenous knowledge can mitigate the impacts of climate change" (Ife 2016). Ultimately, this chapter invites social workers, environmentalists, policy makers, and scholars to rethink their professional and ethical commitments through an eco-social lens—an approach that fosters collective responsibility and systemic change in the face of planetary challenges (Matthies and Närhi 2017; Rambaree et al. 2023). Thus, it calls for a paradigmatic shift from anthropocentric to eco-social imaginaries, which recognize humans as part of interconnected ecological systems.

2.2 The Concept of Eco-Social Life

Imagine a farmer in India waiting in anticipation of the monsoon rains, or a city dweller breathing in clean air in an urban park. In either case, the person's well-being is directly linked to natural systems. The concept of *eco-social life* emphasizes that human communities and the living environment form one interconnected system. From this view, social well-being and ecosystem health are not separate; each depends on the other. Kraut (2007) observes that "plants, animals, and human beings flourish when conditions are favourable." We also understand that the environment is a critical foundation supporting human life and well-being, because it provides the natural resources and ecosystem services necessary for human survival and development (O'Mahony 2022); if the ecosystems collapse, the foundation of human subsistence also breaks down (Salonen and Konkka 2015). Similarly, Kelemen (2023) argued that environmental crises stem from a flawed relationship between humans and nature, and that eco-social well-being transpires through our everyday, ordinary life. Aligning actions with personal values (e.g., simplicity, respect for nature) is central to meaningful well-being and challenges mainstream, consumption-driven lifestyles. This is perhaps embodied in W.H.O.'s *One Health* approach, acknowledging that human, animal and environmental health are all deeply interwoven (EEA 2023).

The concept of biological integrity holds that a system's ability to function relies on maintaining the composition of native species, the processes, and structures (Karr and Dudley 1991). Degradation—especially due to human alteration—reduces this integrity, impairing ecosystem functionality and resilience (Thom and Seidl 2016). Evidence shows that when natural conditions are disrupted by human activity—pollution, habitat disturbance, or climate change—ecosystem services such as carbon storage, nutrient cycling, and water purification, become resultantly impaired (Bowd et al. 2021). Meta-analysis of studies on boreal and temperate forests reveals that disturbances often reduce key services despite sometimes increasing species richness, demonstrating a "disturbance paradox" where natural functions decline (Chapin et al. 2015). Disrupting natural conditions, such as through nutrient pollution or warming, can destabilize complex ecological networks (e.g., plankton communities), weakening interactions and biodiversity that underpin ecosystem functionality (Chapin et al. 2015). All the above evidence seems to imply that human activity disrupts and substantially impedes the functions of natural ecosystems.

2.3 Eco-Social Work

The concept of eco-social work is distinguished from related concepts such as *green social work* and *environmental social work*. While *green social work* emphasizes macro-level green development and justice (Dominelli 2013), *environmental social work* focuses on environmental protection and justice (Ramsay and Boddy 2017). On the other hand, *Eco-social work* integrates indigenous ecological knowledge and cultural diversity into interventions, emphasizing *harmony between humans and nature* and *welfare integration* (Boetto 2017; Matthies et al. 2019). At the same time, eco-social work highlights four extended values which are pertinent: first, *ecological justice and sustainability* wherein it extends justice concerns to nature and future generations (Norton 2012); second, upholding *harmonious coexistence* that moves beyond anthropocentrism, towards human–nature intersubjectivity; thirdly, it pursues *environmental justice and green development* so as to protect marginalized groups from ecological harm caused by industrialism (Dominelli 2013), and finally, eco-social work perpetuates *cultural and ecological empowerment,* enhancing clients' cultural identity and ecological resilience, through participatory practice (Wang and Has 2021; Lysack 2012).

2.4 Eco-Social Work and Systems Theory

Considering that human activity disrupts and substantially impedes the functions of natural ecosystems, social work and allied fields have engaged for a long time with *systems theory* to capture and explain such dynamics. Systems theory is an interdisciplinary approach to complex systems that focuses on "the dynamics and interactions of people in their environments and explains how a system functions and impacts on the person, family, organization, and society" (Flamand 2017). Flamand perceives human beings, families, organizations, and environments as *interacting subsystems* that are open systems, exchanging sources and information with supra- and subsystems over time. Understanding complex social-ecological realities demands engagement with both reductionist (element-level) and holistic (whole-system) lenses.

The *Person-In-Environment* (PIE) approach in social work builds on the above-mentioned concept of *interacting subsystems*, maintaining that individual behaviour "cannot be understood adequately without consideration of the various overlapping and interacting systems" around them. The PIE approach thus provides a more adequate framework for assessing the individual and presenting the problem and strengths, so as to maximize the range of interventions available to the practitioner, with the options to intervene directly with the individual or in aspects of the individual's environment or both (Kondrat 2015). Within an eco-social framework, the *environment* in PIE has been expanded to include ecological systems. As Bellen and Nuria (2024) describe, an eco-social worldview sees "life as fundamentally one, so that humans are inseparable from nature." In practical terms, this means that when a social worker assesses a person's well-being, she/he is bound to consider the state of soils, water, climate, plants and animals that co-exist and affect them. Global social work bodies echo this broad view: for instance, the IFSW states explicitly that "humans are part of the ecosystem, and that human and environmental well-being are interrelated" (IFSW 2022).

Put simply, eco-social life is about mutual dependence. Ecosystems provide the essentials for life, food, water, clean air, and climate regulation, so human societies can only prosper if those systems are healthy. Conversely, human activities can also nurture or degrade ecosystems. An eco-social perspective insists that we care for nature, not only for its own sake, but because it underpins human well-being. In this light, sustainable development, social justice, and disaster management must treat the social and natural worlds as one integrated reality (IFSW 2022; EEA 2023). In fact, *the Earth Charter* (2000) and *UN Sustainable Development Goals* (2015) also emphasized justice, peace, and ecological integrity long ago (Wadcock 2024). Thus, if one starts from the premise of focusing on human well-being, this inclusive approach can still remain grounded in an anthropocentric perspective.

2.5 Planetary Thresholds and Eco-Social Balance

Human society relies heavily on Earth's natural systems. Just as nature sustains human society, humanity profoundly affects as well as alters nature, often with unintended, yet far-reaching consequences. The advent of the industrial revolution and subsequent technological advances have transformed man into a "major geological force" to reckon with, altering land, oceans, the air and climate. Today *climate change, biodiversity loss, pollution and resource depletion* are largely driven by human action: mining oil and minerals, burning fossil fuels, clear-cutting forests, depleting the oceans, intensifying agriculture, overfishing and producing and infusing vast waste and emissions into air, land and water (Flores et al. 2024; Stockholm Resilience Centre 2023).

For a fact, experts argue that we have entered "anthropocene," an epoch defined by human-driven changes to Earth's geology and ecosystems, on a scale both dramatic and comparable to major past geological shifts (Vaughan 2016). Scientists warn of *environmental tipping points*—critical thresholds beyond which any incremental disturbance can trigger abrupt, irreversible climate events (IPCC 2021), drastically affecting the quality of human life (Lenton et al. 2008). Likewise, studies of Earth's "planetary thresholds" indicate humanity is already pushing several safe limits, significantly raising the risk of abrupt, system-wide environmental change. The crossing of these thresholds could cascade into rapid ice-sheet melt, forest die-back or other non-linear shifts that feed back into climate change and biodiversity loss, threatening global food supplies and economies and destabilizing societies worldwide (Stockholm Resilience Centre 2023).

The concept of *planetary boundaries* evolved by the Stockholm Resilience Centre (SRC), provides a powerful framework for understanding the ecological limits within which human societies must operate to maintain a stable Earth system. Introduced in 2009, this framework identified nine critical processes: (1) climate change; (2) biosphere integrity (formerly biodiversity, focusing on loss and function); (3) land-system change; (4) bio-geo-chemical flows (nitrogen and phosphorus cycles); (5) freshwater change/freshwater use; (6) ocean acidification; (7)

stratospheric ozone depletion; (8) atmospheric aerosol loading; and (9) introduction of novel entities, regulating the stability and resilience of the Earth. Various internal papers of the Stockholm Resilience Centre detail the *planetary boundaries* processes assessed and crossed in 2009, 2015, and 2023.

According to the SRC status report in 2009, among the 9 boundary parameters identified, three, climate change, biosphere integrity (Biodiversity Loss) and biogeochemical flows, had been transgressed, with all other boundaries assessed as approaching risk, but not yet exceeded (Rockström et al. 2009). By 2015, four of the nine parameters, climate change, biosphere integrity, land-system change and biogeochemical flows, had crossed the boundaries, while other boundaries were still under or near threshold, or not yet quantified well enough for cross-boundary conclusion (e.g., freshwater use, novel entities, etc.) (Steffen et al. 2015). These findings marked a crucial moment in sustainability debates, as they highlighted how patterns of deforestation, urbanization, and agricultural expansion were not only transforming landscapes but also weakening the biosphere's ability to support social systems. The SRC 2023 status report painted an alarming picture; with all nine boundaries for the first time being assessed comprehensively, with the exception of three, ocean acidification, stratospheric ozone depletion (slightly recovered) and atmospheric aerosol loading (exceeded regionally but not globally), the remaining six had fully crossed the boundaries on a planetary scale (Richardson et al. 2023). In short, while nature sustains us, the collective impact of human activity is now pushing the planet toward critical tipping points, transitions that, once passed, could trigger sudden and irreversible changes that would damage life-support systems.

For an eco-social perspective, these findings underscore the *reciprocal relationship between social and natural systems*. When ecological thresholds are crossed, social risks multiply—through disasters, resource conflicts, and public health crises. Conversely, social policies, technologies, and value systems directly determine whether humanity can remain within these safe ecological limits. The planetary boundaries framework thus strengthens the argument that sustainable social development is impossible without ecological stewardship, and that ecological stability itself depends on just and resilient human systems.

2.6 Man Impacting Nature

The following are the key impacts of human activity (IPCC 2022; IPBES 2019; USEPA 2025):

- *Climate change*: The release of CO_2 and other greenhouse gases from industry, power generation, and deforestation has warmed the planet. The IPCC reports with "very high confidence" that human-caused climate change has already "altered marine, terrestrial and freshwater ecosystems all around the world." Observed impacts include local species extinctions, mass die-offs of plants and animals, and declines in ecosystem services (e.g., pollination, water purification).

Every increase in warming threatens more habitats and species for instance, one IPCC (2022) scenario projects that at a 1.5–2 °C warming, roughly 1 in 10 species in tropical ecosystems could be forced into extinction.
- *Biodiversity loss*: Human consumption of natural resources is unprecedented. The IPBES (2019) reports that "nature is currently supplying more materials than ever before," but at the expense of "unprecedented global declines in the extent and integrity of ecosystems." Such erosion of forests, wetlands, reefs and other habitats reduces the benefits people receive, food, medicines, climate regulation, and "threatens medium-term business continuity and the quality of life for people and communities" worldwide. Without urgent changes, critical ecosystem functions may collapse.
- *Land and resource use*: Expanding agriculture, urban development, mining and dams, fragment and destroy habitats. Alternatively, elsewhere, wetlands are drained dry, jungles logged, and soil eroded. For example, vast rainforest areas have been cleared for cattle ranching or palm oil, releasing carbon dioxide and extinguishing biodiversity. Even if data are scarce, it is clear that conversion of land for human use has outpaced nature's ability to regenerate in many regions.
- *Pollution and waste*: Industrial and chemical by-products contaminate air, water and soil. Plastics accumulate in oceans, endocrine disruptors wash into rivers, and untreated sewage or runoff create dead zones. These pollutants silently poison wildlife and people alike in the long term. The vulnerable are the most affected.
- *Overexploitation*: Overfishing, poaching, and unsustainable hunting drive many species toward extinction. Aquifers are depleted faster than replenished. The "Tragedy of the Commons" plays out as common goods (like fisheries and air) are overused.

Together, these pressures have generated a *"triple planetary crisis"* burdened by *climate change, biodiversity loss* and *pollution*, all human-induced (USEPA 2025). The European Environment Agency (EEA) warns that these interconnected crises call for transformative action to safeguard both nature and society (EEA 2023). Indeed, climate-driven changes have already begun to reshape economies and cultures—from retreating glaciers in the Himalayas to bleaching coral reefs in the Barrier Reefs as well as in Southeast Asia. If unchecked, human impacts on nature will undermine the very life-support systems on which human societies depend.

The impact of human activity carries further ramifications. Mukherjee (Mukerjee 1930, 1942) acknowledges the reciprocal relationship between human settlements and their environments, avoiding ecological determinism while recognizing interdependence. He argued that ecology shapes society, but human intervention also reshapes ecology in dynamic ways. For him, ecological planning was inseparable from social and economic planning, with culture playing a critical role in shaping civic life and maintaining ecological harmony. He classified Indian towns (e.g., capital cities, fortress towns, trading outposts, university towns) based on ecological and cultural functions, demonstrating the integration of ecology with social organization and regional planning (Majumder 2014).

2.7 Nature-Based Solutions (NBS) and Disaster Management for Human Survival

Beyond parks, city planners in more recent times are turning to nature-based infrastructure (NBS), e.g., restoring river floodplains or urban wetlands, to manage floods. A growing body of case studies shows how NBS often outperform traditional "gray" engineering in cost-effectiveness.

- Europe: A Belgian study compared a restored floodplain (an NBS) to conventional stormwater basins and found *similar flood protection* at lower cost, with much higher biodiversity and ecosystem co-benefits under the NBS option (Turkelboom et al. 2021). In this specific instance, in the Dijle valley project intended to reconnect the river Dijle, south of Leuven, central Belgium, to its alluvial floodplain, the river-floodplain restoration delivered the same flood security as technical basins but with *lower construction and maintenance costs*, plus benefits like habitat restoration. More broadly, economic analyses find that NBS cost far less than concrete solutions. A 2022 survey of global case studies reports that nature-based infrastructure delivering a given service is "up to 50% cheaper than grey infrastructure" providing the same service (Bechauf et al. 2022). In other words, building and maintaining green flood controls (like wetlands or green roofs) often requires far lower investment for comparable flood-risk reduction. The same report vouches that NBS yield *28% better value for money* on average, once one counts added health and ecological benefits, especially when co-benefits (recreation, biodiversity, water quality) are included.
- *Asia*: Apparently, in Asia, indigenous wisdom which are nature-oriented contributes to savings. Empirical field research by Badola and Hussain (2005) conducted in Bhitarkanika, comparing villages protected by mangroves, an engineered embankment, or no protection, found that mangrove-covered villages experienced the lowest cyclone damage, with average household losses of US $33.31, compared to US $153.74 in embankment areas and US $44.02 in unprotected areas. Besides, locals were aware and appreciative of mangroves' protective role and actively supported restoration efforts. Tyagi (2022) analysed how the degradation of mangroves in the Mithi River catchment (south Mumbai) contributed to urban flooding. Using hydrologic modelling and flood mapping, Tyagi showed that mangrove loss increased flood inundation area by 1.4% between 1973 and 2005, and peak water depths rose near mangrove-depleted zones—making a case for mangrove restoration as a cost-effective tool for flood mitigation. George et al. (2023) explored NBS interventions in Kochi, Kerala—a tropical coastal city prone to flooding. Solutions included restoring canal networks (e.g., Mullassery Canal), preserving wetlands, planting urban trees, and installing permeable pavements. These approaches were found to significantly enhance flood resilience, improve biodiversity, and align with Kochi's 2040 urban master plan.

- *Globally*: Pelckmans et al. (2024) reviewed global research on how mangroves reduce flood and storm surge risks—specifically in river deltas vulnerable to ENSO events. The paper highlights *significant flood mitigation potential*, advocating for coastal wetland.

2.8 Nature's Role in Shaping Human Lives

In turn, the natural world strongly shapes human destinies—economically, culturally, and emotionally. Ecosystems provide food, water, energy, and raw materials. Geography and climate determine where people settle, what crops they grow, and what hazards they will encounter. We illustrate this point with cases from the Global North (Europe) and the Global South (South and Southeast Asia).

- *Europe (Global North)*: In Europe, nature's imprint is seen in history and policy. Major rivers like the Rhine and Danube, which historically enabled trade and cultural exchange for centuries, while Europe's varied regional climates have dictated agricultural calendars—from vineyards in Southern France to cereal cultivation (wheat) in Northern Europe. In recent decades, heightened flooding and extreme climate events have drawn attention to nature's enduring power. For instance, the catastrophic flooding in Germany during the summer of 2021 prompted the Bavarian government to adopt comprehensive strategies to redesign flood management to accommodate altered Alpine snowmelt regimes and river flow patterns resulting from warming trends (Willkofer et al. 2023). The EEA (2023) observed that nature protection is "important for ensuring healthy ecosystems, a sustainable economy and human well-being," thereby acknowledging that any loss of Europe's forests, wetlands or pollinators would directly undermine European livelihoods and health. In June 2024, the European Union elevated the protection and restoration of ecosystems to the core of its climate agenda, formally adopting its *Nature Restoration Regulation*, committing Member States to revitalize and restore at least *30% of degraded habitats by 2030*, and eventually to *90% by 2050* (European Commission 2024). This landmark law, embedded within the broader *European Green Deal* and the *EU Biodiversity Strategy for 2030*, aims to align biodiversity preservation, climate resilience, and sustainable economic well-being. Invariably, individuals living in urban spaces in the E.U. are becoming more conscious about the benefits of nature; apart from physical fitness, green spaces (parks, riverside paths) improve mental health and social cohesion (EEA 2021), and nature-based solutions (like riverbed restoration) are seen as cheaper, sustainable and eco-friendlier than gray infrastructure for effective and sustainable flood control (EEA 2015).
- *South Asia (India and neighbouring countries):* In South Asia, human lives are deeply entwined with monsoons, rivers, mountains and forests. Agriculture (the livelihood of hundreds of millions) depends on seasonal rains. When the rains flood fields, they replenish soil; yet, when they fail or become erratic, they ruin

harvests (Pathak 2023). In June 2023, for instance, unprecedented heavy monsoon rains inundated parts of Northern India, including Delhi, displacing thousands and killing scores of people. Experts warn that "human-caused climate change is making rain more extreme" and that Himalayan states should expect more "unpredictable and heavy seasons" (Arasu 2023). Himalayan glaciers are a huge lifeline: they operate as the "critical water store" for about 250 million people in the region, feeding the rivers, Indus, Ganges, Brahmaputra, etc., and another 1.65 billion downstream (Carrington 2019). If the glaciers shrink, those rivers and the agriculture, cities, and cultures they sustain, will face a crisis. Coastal South Asia faces other challenges: Bangladesh, a delta nation, exemplifies vulnerability. Approximately 80% of Bangladesh's land is floodplain and nearly two-thirds are situated less than 5 m above sea level (Dasgupta et al. 2014), so storm surges and sea-level rise can easily submerge huge areas. Historically, cyclones like Sidr (2007) and Aila (2009) have devastated millions of lives and destroyed housing and critical infrastructure (Dasgupta et al. 2014). These environmental factors shape the way South Asian society lives: millions adapt their housing, planting schedules, and livelihood strategies to the ebb and flow of nature, often at great human cost. Any kind of disruption in the weather conditions will upset this delicate rhythm.

Besides, there is growing recognition of the importance of greenery, the world over; a recent systematic review found that neighbourhood greenery had "substantial protective effects on the mental health of disadvantaged groups" (Xian et al. 2024). In other words, evidence suggests people living near parks or riversides experienced significantly better mental well-being, even after controlling for other factors. Mechanistically, this benefit often translates through social pathways: various mixed country reviews suggest that urban parks and green spaces improve mental health "mainly through increased social cohesion" by building ties between diverse groups among city dwellers (Haq et al. 2021; Xian et al. 2024).

- *Southeast Asia (e.g., Mekong region, island nations):* Across Southeast Asia, humans have similarly adapted around nature's rhythms. The Mekong River Basin, crossing China, Myanmar, Laos, Thailand, Cambodia and Vietnam, is a prime example. Its seasonal floods nourish rice paddies and fisheries that feed tens of millions. Any change (from upstream dams or changing rainfall) directly affects food security. Coastal communities rely on mangroves and coral reefs for storm protection and fish habitats. For instance, recurring typhoons and rising seas in the Philippines or Vietnam force migrations and alter livelihoods year to year. As a whole, Asia is extraordinarily sensitive to nature's variability: experts note that Asia has high reliance on climate-sensitive livelihoods and high poverty and many people living in vulnerable areas (Krishnamurthy et al. 2023). Thus, environmental changes, from delayed monsoons to salinity intrusion, cascade, precipitating social crises of hunger, disease, or displacement.

In every region, nature also provides intangible benefits that shape cultures: sacred rivers, revered mountains, spiritual beliefs about forests or animals. For example, India's Ganges River is held sacred and embedded in social life; in Andean cultures, Pachamama (Mother Earth) is a living entity. These cultural ties reflect how societies *construct* their identities as well as their sense of connection with nature (refer to the next section). What is clear is that when ecosystems change, whether through climate, pollution or resource use, human lives, too, change. Food, water, health and shelter all depend on ecological balance. As one analysis puts it, the ongoing loss of nature's integrity "will threaten livelihoods and food security" globally (IPCC 2022). In sum, the material and cultural foundations of life, from agriculture to livelihood, public health to spirituality, are all moulded by nature's gifts as well as constraints.

2.9 How Societies Culturally and Historically Construct Their Relationship with Nature

Beyond the material links, the *meaning* of nature is often shaped by culture and history (Evenden 2013). Societies do not universally view nature in the same way: many industrial societies (the West) historically treated nature as a resource to be mastered and optimized, using science and technology to exert control. During the Middle Ages and beyond within the Western Christian tradition, a dominant interpretation of Genesis taught that humans, created in God's image, have dominion over creation, justifying an anthropocentric, utilitarian approach towards Nature (White 1967; University of Exeter n.d.), legitimising exploitation and reduced "animistic reverence" toward nature (Dominion Theology). This mindset underpinned rapid development post-Renaissance, but unfortunately, in hindsight, at the cost of ecological imbalance. In contrast, eco-theological thinkers urge a shift toward "stewardship," emphasizing care, interdependence, and ethical responsibility for creation (Schaefer 2023; Bugg 1991). This outlook echoed in movements rooted in integral ecology, such as Pope Francis's "Laudato si," which frames environmental protection as a spiritual and moral imperative grounded in gratitude for Earth as God's gift (Francis 2015; Schaefer 2023). This is quite in line with the approach and practice of many indigenous and non-Western cultures, which perceive humans as part of nature or kin to natural beings. Animist or earth-based traditions from tribal societies in Asia to Native American or Pacific Islander beliefs, emphasize respect, reciprocity and the sacredness of ecosystems.

This cultural framing affects how people behave. The 1972 essay "Should Trees Have Standing?" by Christopher D. Stone (2010) laid the philosophical foundation proposing environmental personhood for trees, rivers, and air. So have there been legal discourses suggesting "valleys, alpine meadows... rivers... groves of trees... be granted legal personhood"? (Mohd Salim v. State of Uttarakhand and Others 2017; Sierra Club v. Morton 1972). In Ecuador, for instance, indigenous cosmology influenced national law: the 2008 constitution (reflecting *pachamama* beliefs) grants nature legal rights. It declares that "Nature, or *pachamama*... has the right to integral respect for its existence and for the maintenance and regeneration of its life cycles" (Challe 2021). By legally centering nature, Ecuador challenged the notion of nature as mere property. Similarly, social movements worldwide (e.g., "Rights of Nature" campaigns) rest on non-traditional worldviews (Posner 2021). Even in the Global North, changing ideas are seen in policy; the EU's biodiversity strategy recognizes "nature's intrinsic value" alongside its economic utility, signalling a paradigm cultural shift.

Scholars in *ecocriticism* analyse exactly these narratives. They examine literature, media, and political discourse to see how societies construct their own indigenous narratives on nature. For instance, one review notes that ecocritical perspectives highlight *"the structural causes of the development of modernity, which break the balance of biophysical ecosystems,"* and suggest alternative models like degrowth and "good living" (D'Alisa et al. 2014). In other words, ecocriticism asks us to question the taken-for-granted idea that endless economic growth is good, pointing out instead the losses inflicted on natural systems.

Culture also determines whose voice counts in managing nature. Often, marginalized communities (indigenous peoples, ethnic minorities, the poor) bear the worst environmental burdens. As Bellen and Nuria (2024) emphasize, the "most vulnerable communities are most affected by environmental risks and their effects are amplified" for children, the elderly, women, and the economically poor. For example, a chemical disaster and the resultant pollution on the scale of the Union Carbide Bhopal Tragedy or a cyclone or a hurricane on the Category 5 Katrina may hit a slum, the coastal or tribal village hardest, because people there have far fewer resources to adapt. In some cases, such as Hurricane Katrina in 2005, this can even go so far that no one cares for the bodies (see Illner 2020, Chap. 1: "2005 – The Unclaimed Corpses). Such unfortunate events reveal how environmental justice is deeply embedded in cultural and power structures, where socially constructed values determine who is protected and who is exposed—who is deemed "worthy" and who is treated as "expendable." Thus, "environmental justice" issues reflect cultural and power structures. Socially constructed values, who is "worthy" and who is "expendable," deeply shape the human-nature relationship.

In sum, how a society *conceptualizes* nature is historically contingent. Some societies tell stories of dominion over nature; others celebrate stories of kinship. These narratives influence policy, ethics, and everyday choices. If nature is seen primarily as "resource," policies will prioritize short-term exploitation as envisaged in the anthropocentric "*frontier ethic*"; early European settlers in North America rapidly, having consumed and depleted the natural resources of the land, moved westward to conquer new frontiers (Fisher 2022). Frontier ethic was hinged on the assumption that there existed infinite resources to indefinitely support continued population growth. But now that the population is fast approaching 8.5 billion and with few frontiers left, many have begun questioning the frontier ethic, and are moving toward an *environmental ethic*, which includes humans as part of the natural community, rather than managers of it. Such an ethic places limits on human activities (e.g., uncontrolled resource use), that may adversely affect the natural community. Expanding on this, alternatively if nature is seen as "community member" (as indigenous people do), policies might focus on care, stewardship and preservation. Social workers and development professionals must therefore distinguish between these cultural constructions. An ecocritical lens in eco-social work urges practitioners to uncover the hidden assumptions in dominant ideologies such as the consumerist belief that "more is better," and the ethical community approach that prioritises coexistence and the optimal use of resources, as exemplified in sustainable development. The latter approach is not inherently "anti-consumerist" either. The latter positions have been upheld by indigenous populations, e.g., by First Nation Custodial Ethics (Australia) or Kaitiakitanga (Māori concept of guardianship and care for the environment) (Marsden and Henare 1992; Peters 2017). Only by understanding these worldviews can societies rewrite their human nature story and direct it more toward justice and sustainability.

2.10 Risk Society, Modernity's Paradox and Role of Social Work (Intervention)

Modern technological society has generated new kinds of risks that transcend geography and social class. Ulrich Beck (1992) famously termed our era a *risk society* (Risikogesellschaft) wherein *manufactured dangers*, nuclear radiation, climate change, and pandemics, emerge from the very systems designed to advance human progress; and industrialisation creates global hazards that no single nation can escape unscathed, as witnessed during the Chernobyl nuclear disaster (1986) or Covid-19. In a risk society, dangers do not come from lacking modernity (like traditional epidemics) but from *too much* modernity, so to speak. Climate change is a classic case: it is a product of human industry and now threatens humanity. As Lübken explains, these "manufactured" risks are pervasive, yet unequally distributed. Climate change "does not threaten all individuals or nations equally," he observes (Culver et al. 2011); wealthier countries and populations have more resources to adapt or retreat. In other words, although greenhouse gases mix globally, the *impacts* of climate hazards are culturally and economically mediated.

To be read in tandem is Edward Lorenz's *butterfly effect*—a cornerstone of *chaos theory*, demonstrating how minor perturbations (figuratively, a butterfly flapping its wings) can dramatically alter large systems like weather. In modernity, innovations, such as cars, factories, and chemical production, though life-enhancing, have inadvertently destabilized planetary systems, triggering climate change, biodiversity loss, and health risks. These unintended consequences illustrate how seemingly small technological advancements in one part of the world can cascade into global environmental crises—truly a modern "butterfly effect," which seems central to modernity; the very advances (cars, factories, chemicals) that improved lives have eventually also destabilized planetary systems, bringing in devastating consequences. Yet modern societies still depend on these systems.

Beck would argue that recognizing these risks, prevention and counteraction requires "cosmopolitan" cooperation (global agreements and local action). In practice, this implies that global institutions such as the UNFCCC, the local self-governments and local disaster management must work together. It also means questioning blind faith in technology. As Gray and Coates ask: if humans caused the ecological crisis, must humanity not also be the solution? They note that *human intervention* is both the cause of environmental problems and the means to restore balance (Bellen and Nuria 2024).

For social work and disaster risk reduction, *risk society* highlights the need for systemic thinking and preparedness. Such an understanding reinforces the fact that we cannot rely solely on expertise or charity; instead, we must engage communities as active agents. This is where a *strengths-based approach* (SBA) becomes vital. Rather than seeing communities only as victims of climate extremes, a strengths perspective identifies what *is working* in a community, local knowledge, traditions, social networks, and builds on it in handling, preventing, mitigating and managing disasters. Research in disaster contexts shows the power of such an orientation: for example, community-driven child protection programs in war-affected areas achieved "high levels of community ownership and effective, sustainable supports" for vulnerable children by leveraging local resources (Wessells 2016; Tol et al. 2013). By analogy, climate adaptation programs that empower people, giving them tools, land rights, and recognition of their skills, tend to create deeper resilience than top-down aid. In a world of manufactured uncertainty, strengthening peoples' capacities is as important as reducing emissions.

Thus, risk society and modernity force a paradigm shift: from seeing nature as a predictable backdrop to treating it as a dynamic feedback loop with society. Professionals must help communities navigate these complexities. For instance, we teach communities early-warning (for cyclones) and equip them to activate and structure disaster management, we diversify livelihoods (against drought) and also advocate for structural change (like environmental regulation) (Mallick et al. 2014). In all these cases, we pair technical fixes with an appreciation that the *modern* genesis of these risks means we need *modern* transformations in thinking on how to handle them in a more holistic and effective manner.

2.11 Power, Policy, and Eco-Social Change

Addressing eco-social challenges ultimately depends on power relations and policy choices. Economic, political and social power determine who makes decisions about nature and who benefits or suffers. For instance, multinationals and affluent nations have historically dictated fossil fuel use and resource extraction, often sidelining the Global South and marginalized groups. Upper-middle and high-income countries have high levels of material consumption. The 2024 Global Resource Outlook unambiguously details the damage as follows. It documents that the extraction of the earth's natural resources, related to the massive build-up of infrastructure in many parts of the world, has tripled over the past five decades. The extraction and processing of fossil fuels, metals and non-metallic minerals (e.g., sand, gravel, clay) together account for 35% of global emissions. Material extraction is projected to rise by 60% by 2060. The extraction and processing of biomass (e.g., agricultural crops and forestry) accounts for 90% of land-related biodiversity loss, water stress, as well as one-third of greenhouse gas emissions. The burgeoning resource use since 1970 from 30 to 106 billion tonnes, or the consumption of 23–39 kg of materials used on average per person per day, has dramatic environmental impacts. Overall, resource extraction and processing account for over 60% of planet-warming emissions and for 40% of health-related impacts of air pollution. Put together, all these could derail efforts to achieve not only global climate, biodiversity, and pollution targets but also economic prosperity and human well-being, according to a report published today by the UN Environment Programme (UNEP)-hosted International Resource Panel (UNEP 2024). Shifting to a just eco-social future requires rebalancing that power.

International agreements set broad goals, but largely implementation has been painfully uneven; the Paris Agreement and UN Sustainable Development Goals (SDGs) link climate and development, yet experts warn we are not on track. The IPBES Report (2019) explicitly warns that "the current development and environmental trends are incompatible with meeting the SDGs and Paris Agreement goals," and so bluntly calls for urgent action. In response, entities like the European Union launched the Green Deal as a growth strategy toward carbon neutrality. Research underscores that climate and sustainable development objectives are interlinked yet not always aligned. For instance, Laumann et al. (2020) demonstrated complex, non-linear interactions between climate goals and various SDGs, highlighting potential trade-offs and synergies. A machine-learning text analysis of the 74 policy documents underpinning the EU Green Deal (2019–2023) found strong alignment with environmental SDGs: SDG 7 (clean energy), SDG 12 (responsible consumption), and SDG 13 (climate action). However, the SDG Tracker highlights the underrepresentation in social goals: SDG 1 (no poverty), SDG 2 (zero hunger), SDG 3 (health), SDG 4 (education), SDG 5 (gender equality), SDG 8 (decent work), SDG 10 (inequality), SDG 16 (peace and strong institutions), suggesting that SDG progress tracker highlights that social goals like reducing inequality (SDG 10) and a lag behind environmental objectives (Koundouri et al. 2024). This gap illustrates how policy can still prioritize ecology over people, or vice versa.

At the same time, non-governmental movements are pushing back. Grassroots environmental justice campaigns, youth climate strikes, and innovations like "rights of nature" laws are changing discourse (Posner 2021). For example, after floods and landslides, Indian courts and activists have demanded stricter environmental regulations and accountability by repeatedly forcing the Governments at the centre and state levels through court intercession. The Gadgil Committee (Western Ghats Ecology Expert Panel) and Kasthurirangan Committee (High-Level Working Group) Reports pertaining to the Western Ghats, a biodiversity hotspot in Southern India, may be seen as attempts to bring to account the respective governments. Both these reports recommended that demarcation of 37–64% of the Western Ghats be designated as Ecologically Sensitive Areas (ESAs) with strict limitations on development, aiming to protect critical zones while allowing for sustainable development in other (Gadgil 2011; Hindustan Times 2014). In Ecuador, indigenous activists helped enshrine nature's rights in the constitution (Challe 2021). These shifts show society moving toward more eco-centric governance, though unevenly.

More importantly, this is also a call for the social sector to engage. The IFSW's "eco-social approach" outlines how social workers should act on these issues. It stresses that social policy cannot ignore environment: "social justice cannot be fully achieved without taking into account ecological justice" (Bellen and Nuria 2024). In practice, this means social workers are bound to advocate for environmental health as a form of human rights. For instance, co-building an eco-social world, as IFSW phrases it, requires that climate action be *inclusive* and respectful of cultural identities (IFSW 2022). Policies thus should not pit social gains against environmental protection, but seek synergies (e.g., green jobs, urban parks in poor neighbourhoods, community-led reforestation).

A *strengths-based approach* fits neatly here. In disaster risk reduction, practitioners increasingly emphasize community assets and agency. Programs in places like Nepal or Indonesia build on local coping mechanisms (flood-resilient housing, communal savings) rather than imposing "expert," outside solutions. Research confirms this: community-driven action based on local strengths tends to be more sustainable and empowering (Wessells 2016). In climate policy, this means listening to vulnerable communities (e.g., indigenous or coastal villages) as active partners, not passive victims, besides engaging with them. It also means amplifying success stories, such as farmer cooperatives adopting agro-forestry, so they can inform broader change.

Arguably and quite logically there are divergent views. There is an argument that climate change mitigation and adaptation should not be put in one bowl. In Germany, this view is sometimes seen as a kind of trade-off when it is argued that the money at hand should go into adaptation instead of mitigation, given that "climate change will happen anyway" (IPCC 2021). By this argument it is so much better to be prepared against it instead of wasting money on trying to prevent something that cannot be prevented anyway. However, this framing can be problematic. This framing is problematic in that (Lutz et al. 2025; RAND Corporation 2024). It overlooks that adaptation alone does not reduce the rate or scale of climate impacts. As warming intensifies, adaptation costs escalate—often exponentially. (2) It misrepresents the economic case: adaptation investments can offer high returns, but they need not necessarily substitute for the global public good of mitigation. For instance, investing \$1B in coastal adaptation can reduce \$14B in future flood damages—but these measures will need constant upgrading if emissions persist or accelerate, ultimately becoming unviable without mitigation. (3) Economics and global public goods logic: adaptation tends to benefit specific regions or communities immediately, whereas mitigation yields shared global benefits—emission reductions anywhere help everyone. (4) Integration is valuable, but not always practical. Many climate, finance and development practitioners support combining efforts. However, surveys show that over two-thirds of experts see key challenges in full integration: dilution of focus, administrative complexity, competition for funding streams, divergent metrics and priorities (carbon vs vulnerability, efficiency vs equity).

Ultimately, eco-social change will require transforming our collective worldview. Scholars like Boetto and Bowles argue for a *Transformative Eco-social Model* in social work: abandoning the modernist, consumerist mindset and adopting a holistic, ecological one (Boetto 2017). They poignantly write that transformative change "involves a movement away from contemporary consumerist lifestyles… towards lifestyles focused on sustainability and harmony with the natural environment." In other words, policy and culture must promote ecological values and recognize humans as part of nature. This is reflected even in ethics and education: some propose that we should conceive of autonomy as "ecological autonomy" (Chackal 2018) and that educators should aim for "eco-social autonomy" (Pulkki and Keto 2022). Likewise, it can be argued that social workers have a duty to care for all life, not only humans, extending the "person-in-environment" idea to all beings.

Power shifts are visible in some places. A range of stakeholders, governments, businesses, NGOs, and citizens, are (slowly) collaborating. The EEA notes that an effective transition requires "government action as well as business and individual engagement" to rebuild our connection with nature (EEA 2023). The social sector can facilitate this engagement. For example, community development programs often incorporate environmental stewardship, urban gardening projects improve food security while connecting people to soil, and youth groups' tree-planting fosters hope and collective efficacy.

As a field, social work is just beginning to embrace its role. Some theorists insist that caring for Earth must become part of social care. Besthorn and Saleebey provocatively state that caring for people includes caring for the planet—a *post-human* ethics of social work (Bellen and Nuria 2024). Even without those words, the message is clear: social work cannot truly promote well-being while ignoring ecology.

2.12 Discussion and Critique on Systems Theory

All said and done, in retrospect, social work cannot afford to be naïve about the hype regarding the advantage of integrating ecology and social life within the systems perspective. Alternately, it needs to accommodate the divergent and sometimes tangential perspectives regarding the theories incorporated. In this particular instance, although systems theory advocates for integrated, reciprocal feedback across subsystems, many modern business organizations, especially *technocratic, business-driven entities*, systematically violate these principles concerning sustainability and carbon footprint:

1. *Technocratic bias*: By prioritizing technical solutions and *expert-driven decision-making*, these organisations tend to marginalize social, ethical, and environmental perspectives. Driven by profiteering and cajoled by the boards, they are forced

to narrow the focus on technological fixes rather than holistic systemic change (Boyer 2020; Bronfenbrenner (2009).

2. *"Greenwashing"*: Big-tech and large corporations often promote and celebrate "sustainability metrics" tied to voluntary or self-reported targets (e.g., carbon neutral claims via Renewable Energy Certificates [REC] purchase), while opaque feedback mechanisms conceal the true environmental impact—violating the transparency essential for open-system integrity (Ma and Duan 2024). An investigation of corporate sustainability pledges shows that companies frequently use carbon credits and accounting tricks to meet net-zero targets without cutting real emissions. Bernal (2020) notes that many firms have "opted instead to purchase 'carbon credits'" while changing little of their operations and cites an EU report casting doubt on whether such offsets "are making any difference whatsoever" in reducing emissions. For e.g., the Big Three, Microsoft, Amazon, and Meta, use RECs in reports, but objective investigations reveal their actual emissions are in the range of 3.3–8.5 M metric tons, higher than reported (Hern 2024).
3. *Absence of accountability and transparency on "sustainability practices"*: Many corporates rely heavily on "self-reported" data, with limited verification and clarity, undermining the credibility of "green" claims made. Even when restoration or emission reduction claims are made, many provide no cost data, and numerous others omit critical details such as project size or monitoring methodology—a strong indicator of incomplete transparency (New Climate Institute and Carbon Market Watch 2022). A recent academic analysis by Vrikki (2024) of big tech companies' sustainability reports finds that the sector heavily depends on renewables and offsets, raising concerns of greenwashing. It notes, for instance, that Apple, Google, Amazon, Meta and Microsoft boast of sourcing renewable electricity and buying carbon offsets, but whether these measures "are truly effective in mitigating emissions, or if they merely serve as a symbolic gesture without delivering genuine environmental impact" remains questionable. In other words, their use of offsets/RECs can mask ongoing emissions rather than eliminate them, explicitly linking such accounting practices to greenwashing (Vrikki 2024). Meng (2024) cites a news report highlighting how major AI-driving tech firms such as Amazon, Microsoft and Meta obscure or "conceal their actual carbon footprints" by purchasing so-called unbundled renewable energy certificates (RECs). These credits make it appear "that power consumed from a coal plant came from a solar farm instead" allowing these companies to claim huge emission reductions on paper. Hurst and Baazil (2021) cite a Netherlands watchdog's finding that Shell's "drive CO_2 neutral" campaign was misleading, since Shell "could not prove the scheme fully offset emissions." The same was confirmed by a Greenpeace audit, which found Shell's tree-planting projects (offsets) offered "no clear benefit to the climate."
4. *Neglected interdependencies*: A systems approach entails recognizing systemic interconnections and feedback loops. Yet technocratic firms frequently ignore, to a great extent, or even white-wash critical *social and ecological feedback*, resulting in unintended effects such as resource rebound (e.g., greater efficiency may

spur increased consumption) and lock-in within unsustainable extractive practices.

5. *Excluding broader subsystems*: These companies, driven by profit-motive and accountability to the shareholder, typically exclude wider community and ecological stakeholders from decision-making, thus lacking the rich multi-directional inter-system communication that systems theory demands.

2.13 Conclusion

In summary, it is clear that eco-social life is not a static equilibrium, but rather a dynamic, ever-evolving co-creation between human societies and the ecosystems they inhabit. Socio-ecological symbiosis is best understood as an ongoing process of co-evolution, adaptation, and resilience-building, not a fixed state to be attained. Symbiotic sustainability invites us to envision human systems that do not merely minimize harm, but actively nurture regenerative feedback loops, where ecosystems and social well-being are mutually reinforcing. Achieving this vision requires structural, social and cultural transformation—valuing natural capital, embracing circularity, and embedding social equity into every decision. It requires inclusive governance, participatory stewardship, and the integration of diverse knowledge systems—from indigenous ecological wisdom to scientific research and community innovation. This is not just about environmental idealism: it is essential for building resilient, equitable futures capable of withstanding global change. Ultimately, the promise of eco-social life lies in recognizing our interdependence with nature, designing our economies, institutions, and cultures accordingly. When we shift from competitive extraction to collaborative, regenerative relationship, we move toward a world where both human and ecological communities can thrive together.

However, the balance between man and nature is a thin line and remains dilemmatic. For anyone trying to come to terms with the contradictions marking our efforts to live sustainably by balancing between the social world and the natural world, we cannot oversimplify or underestimate the force of the wild, untamed nature. For the social worker with all good intentions to engage and defend nature, one needs to consider the dilemma posed to those who choose to devote themselves entirely to values and idealism. Environmentalism becomes for a rare few a lifelong commitment, for some a passing experiment, and for others what they strategically mobilize to change their own pollution-ridden environment into something more habitable. In many cases, that implies sacrificing personal relationships and community ties, or even the dilemma of translating their lofty goals into decisive collective action. Mathias (2024) expands on this dilemma, drawing a clear distinction between "living for" (a goal or cause), and "living from" (a particular cultural context) to ponder not just the various pressures that constitute contemporary activism in the context of Kerala (southwestern India). Thus, *eco-social life* is a very complex paradigm for a practitioner, whether it be a social worker, a policy-maker, or a social engineer, considering that the tensions between community and cause, family and purpose, living *for* and living *from* and the various dilemmas it presents by way of its ethics, its hopes, its failures, with the growing urgency of environmental injustice with each passing day.

Eco-social change is inherently about power and policy. It debates: Who controls resources? Who speaks for nature? What laws protect or endanger life-support systems? By using systems theory and PIE, professionals can see how local issues (poverty, housing, health) tie into global ecology. An ecocritical lens keeps us aware of hidden assumptions in the status quo. A strengths-based approach ensures that vulnerable communities are partners in change, not casualties. And transformative eco-social models remind us to reimagine our role on Earth. As Bellen and Nuria (2024) concludes, safeguarding *eco-social well-being* means expanding our ethical community to "include soil, water, plants and all living things." In doing so, we take a holistic step toward a future where neither people nor the planet is an afterthought, but a shared life in harmony.

References

Arasu S (2023) India's devastating monsoon season is a sign of things to come, as climate and poor planning combine. Associated Press. https://apnews.com/article/monsoons-climate-change-disasters-india-himalayas-landslides-rains-b1d09581 1f252ab82268dbefbeef5b3a

Badola R, Hussain SA (2005) Valuing ecosystem functions: an empirical study on the storm protection function of Bhitarkanika mangrove ecosystem. India Environ Conserv 32(3):85–92

Bechauf R, Cutler E, Bassi AM, Casier L, Kapetanakis M, Pallaske G, Simmons B (2022) The value of incorporating nature in urban infrastructure planning: IISD report. International Institute for Sustainable Development (IISD). Retrieved from https://nbi.iisd.org/nature-infrastructure-planning

Beck U (1992) Risk society: towards a new modernity (trans: Ritter M). SAGE Publications

Bellen PR, Nuria PB (2024) The responsibilities of social work for ecosocial justice. Soc Sci 13(11):589. https://doi.org/10.3390/socsci13110589

Bernal N (2020) Google, Microsoft and the strange world of corporate greenwashing. WIRED. https://www.wired.com/story/corporate-greenwashing/

Boetto H (2017) A transformative eco-social model: challenging modernist assumptions in social work. Br J Soc Work 47(1):48–67

Bowd B, Blanchard W, McBurney L, Lindenmayer D (2021) Direct and indirect disturbance impacts on forest biodiversity. Ecosphere 12(12):3823. https://doi.org/10.1002/ecs2.3823

Boyer J (2020) Toward an evolutionary and sustainability perspective of the innovation ecosystem: revisiting the Panarchy model. Sustainability 12(8):3232. https://doi.org/10.3390/su12083232

Bronfenbrenner U (2009) Bronfenbrenner's ecological systems theory. In: Encyclopedia of early childhood development. Springer. (Original work published 1979)

Bugg C (1991) Stewardship. In: Holman bible dictionary. Holman, Nashville, pp 1303–1304

Carrington D (2019) A third of Himalayan ice cap doomed, finds report. The Guardian. https://www.theguardian.com/environment/2019/feb/04/a-third-of-himalayan-ice-cap-doomed-finds-shocking-report

Chackal T (2018) Place, community, and the generation of ecological autonomy. Environ Ethics 40(3):215–239. https://doi.org/10.5840/enviroethics201840321

Challe T (2021) The rights of nature – can an ecosystem bear legal rights? State of the Planet. https://news.climate.columbia.edu/2021/04/22/rights-of-nature-lawsuits/

Chapin FS, Sommerkorn M, Robards MD, Hillmer-Pegram K (2015) Ecosystem stewardship: a resilience framework for arctic conservation. Glob Environ Chang 34:207–217

Culver L, Egner H, Gallini S, Kneitz A, Lousley C, Lübken U, Mincyte D, Mom G, Winder G (2011) Revisiting risk society: a conversation with Ulrich Beck. RCC Perspectives, 2011(6). Rachel Carson Center for Environment and Society. https://www.environmentandsociety.org/sites/default/files/2011_6_risk_society.pdf

D'Alisa G, Demaria F, Kallis G (2014) Degrowth: a vocabulary for a new era. Routledge, New York

Dasgupta S, Huq M, Khan ZH, Ahmed MMA, Mukherjee N, Pandey K, Khan MF (2014) Cyclones in a changing climate: the case of Bangladesh. Clim Dev 6(2):96–110. https://doi.org/10.1080/17565529.2013.868335

Dominelli L (2013) Environmental justice at the heart of social work practice: greening the profession. Int J Soc Welf 22(4):431–439

EEA (2015) Exploring nature-based solutions: the role of green infrastructure in mitigating the impacts of weather- and climate change-related natural hazards EEA. Technical report no 12/2015. https://www.eea.europa.eu/publications/exploring-nature-based-solutions-2015

EEA (2021) Who benefits from nature in cities? Social inequalities in access to urban green and blue spaces across Europe. EEA report no 23/2021. https://www.eea.europa.eu/publications/who-benefits-from-nature-in-cities

EEA (2023) Strengthening society's engagement with nature. https://www.eea.europa.eu/en/analysis/publications/strengthening-societys-engagement-with-nature

European Commission (2024) Council adopts Nature Restoration Regulation. https://www.consilium.europa.eu/en/press/press-releases/2024/06/17/nature-restoration-law-council-gives-final-green-light/

Evenden M (2013) Beyond the culture of nature: introduction. J Can Stud 47(3):5–15. https://doi.org/10.1353/jcs.2013.0021

Fisher MR (2022) Frontier ethic. In Environmental ethics and justice (Section 18.01). In: Introduction to environmental science, 2nd edn. Bakersfield College. LibreTexts. Retrieved https://bio.libretexts.org/Courses/Bakersfield_College/Introduction_to_Environmental_Science/18%3A_Environmental_Justice_and_Pollutants/18.01%3A_Environmental_Ethics_and_Justice

Flamand L (2017) Systems theory of social work. People of everyday life. Retrieved from http://peopleof.oureverydaylife.com/systems-theory-social-work-6260.html

Flores BM, Montoya E, Sakschewski B, Nascimento N, Staal A, Betts RA et al (2024) Critical transitions in the Amazon forest system. Nature 626(7999):555–564

Francis (Pope) (2015) Laudato si': on care for our common home [Catholic Church encyclical letter]. Libreria Editrice Vaticana. Retrieved from https://www.vatican.va/content/francesco/en/encyclicals/documents/papa-francesco_20150524_enciclica-laudato-si.html

Gadgil M, Western Ghats Ecology Expert Panel (2011) Report of the Western Ghats ecology expert Panel. Ministry of Environment and Forests, Government of India. Retrieved November 1 2013, from http://moef.nic.in/downloads/public-information/wg-23052012.pdf

George TM, Prasad S et al (2023) Nature-based solutions for flood mitigation: the case study of Kochi. Sustainability 17(5):1983

Hannigan J (2006) Environmental sociology. Routledge, London

Haq SMA, Islam MN, Siddhanta A, Ahmed KJ, Chowdhury MTA (2021) Public perceptions of urban green spaces: convergences and divergences. Front Sustain Cities 3:755313. https://doi.org/10.3389/frsc.2021.755313

Hern A (2024) Data center emissions probably 662% higher than big tech claims. Can it keep up the ruse? The Guardian. Retrieved from https://www.theguardian.com/technology/2024/sep/15/data-center-gas-emissions-tech

Hindustan Times (2014) Why India must protect the Western Ghats from getting lost. Hindustan Times

Hurst L, Baazil D (2021) Dutch ad watchdog tells Shell to pull 'carbon neutral' campaign. Bloomberg, 27 Aug 2021. https://www.bloomberg.com/news/articles/2021-08-27/dutch-ad-watchdog-tells-shell-to-pull-carbon-neutral-campaign

Ife J (2016) Eco-social justice: re-imagining social work. Int Soc Work 59(5):628–639. https://researchoutput.csu.edu.au/files/22629624/9004792_Published_article_OA.pdf

Illner P (2020) Disasters and social reproduction: crisis response between the state and community. Pluto Press, London

Intergovernmental Panel on Climate Change (IPCC) (2021) Climate Change 2021: the physical science basis. Summary for policymakers. Contribution of working group I to the sixth assessment report of the IPCC. Cambridge University Press

Intergovernmental Panel on Climate Change (IPCC) (2022) Biodiversity: fact sheet—AR6 WGII. Intergovernmental Panel on Climate Change. https://www.ipcc.ch/report/ar6/wg2/downloads/outreach/IPCC_AR6_WGII_FactSheet_Biodiversity.pdf

International Federation of Social Workers (2022) The role of social workers in advancing a new eco-social world. https://www.ifsw.org/the-role-of-social-workers-in-advancing-a-new-eco-social-world/

IPBES (2019) Global assessment report on biodiversity and ecosystem services. Intergovernmental Science-Policy Platform on Biodiversity and Ecosystem Services. https://files.ipbes.net/ipbes-web-prod-public-files/webform/impact_tracking_database/46408/2019%20IPBES%20Biodiversity%20and%20Ecosystem%20Services%20report%20Final.pdf

Karr JR, Dudley DR (1991) Ecological perspective on water quality goals. Environ Manag 15(5):711–719

Kelemen P (2023) Eco-social approach to well-being: the search for eco-social well-being in everyday life according to articles from Helsingin Sanomat. Master's thesis, Tampere University of Applied Sciences

Kondrat M (2015) Person-in-environment. Oxford Bibliographies in Social Work. https://doi.org/10.1093/obo/9780195389678-0092

Koundouri P, Alamanos A, Plataniotis A, Stavridis C, Perifanos K, Devves S (2024) Assessing the sustainability of the European green Deal and its interlinkages with the SDGs. npj Climate Action 3:23. https://doi.org/10.1038/s44168-024-00104-6

Kraut R (2007) What is good and why: the ethics of well-being. Harvard University Press, Cambridge

Krishnamurthy PK, Lewis K, Kent C, Aggarwal P (2023) Climate impacts on food security and livelihoods in Asia: a review of existing knowledge. World Food Programme Regional Bureau for Asia; UK Met Office Hadley Centre; International Water Management Institute. https://weadapt.org/wp-content/uploads/2023/05/wfp281745.pdf

Laumann F, von Kügelgen J, Barahona M (2020) Non-linear interlinkages and key objectives amongst the Paris Agreement and the Sustainable Development Goals. arXiv:2004.09318

Lenton TM, Held H, Kriegler E, Hall JW, Lucht W, Rahmstorf S, Schellnhuber HJ (2008) Tipping elements in the earth's climate system. Proc Natl Acad Sci 105(6):1786–1793. https://doi.org/10.1073/pnas.0705414105

Lutz C, Becker L, Kemmler A, Reuschel S, Sander L, Stöver B (2025) Macroeconomic impacts of climate change, climate adaptation, and climate mitigation in Germany: a PANTA RHEI model analysis. Sustainability 17(13):6175. https://doi.org/10.3390/su17136175

Lysack M (2012) Emotion, ethics, and fostering committed environmental citizenship. In: Environmental social work. Routledge, London, pp 231–245

Ma G, Duan M (2024) Double counting of emission reductions undermines the credibility of corporate renewable energy claims. Environ Sci Technol 58:11247–11255. https://doi.org/10.1021/acs.est.4c03792

Majumder TB (2014) Radha Kamal Mukherjee and his idea of ecology and environment. Indian JL Just 5:167

Mallick B, Rahaman KR, Vogt J (2014) Cyclones in a changing climate: the case of Bangladesh. Int J Disaster Risk Reduction 7:68–75. https://www.researchgate.net/publication/262576954

Marsden M, Henare TA (1992) Kaitiakitanga: a definitive introduction to the holistic world view of the Maori. Unpublished manuscript

Mathias J (2024) Uncommon cause: living for environmental justice in Kerala. University of California Press, Oakland. https://doi.org/10.1525/9780520395510
Matthies AL, Närhi K (eds) (2017) The ecosocial transition of societies: the contribution of social work and social policy. Routledge, London
Matthies AL, Stamm I, Hirvilammi T, Närhi K (2019) Ecosocial innovations and their capacity to integrate ecological, economic and social sustainability transition. Sustainability 11(7):2107
Meng M (2024) How tech companies are obscuring AI's real carbon footprint. Luxembourg Times. https://www.luxtimes.lu/world/how-tech-companies-are-obscuring-ai-s-real-carbon-footprint/18204600.html
Mohd Salim v. State of Uttarakhand & Others, Writ Petition (PIL) No. 126 of 2014 (High Ct. Uttarakhand Mar. 20, 2017)
Mukerjee R (1930) The regional balance of man. Am J Sociol 36(3):455–460
Mukerjee R (1942) Mobility, ecological and social. Soc Forces 21:154–159
New Climate Institute & Carbon Market Watch (2022) Corporate Climate Responsibility Monitor 2022. Retrieved from https://newclimate.org/wp-content/uploads/2022/02/CorporateClimateResponsibilityMonitor2022.pdf
Norton B (2012) Solar energy thermal technology. Springer Science & Business Media
O'Mahony T (2022) Toward sustainable wellbeing: advances in contemporary concepts. Front Sustain 3:807984
Pathak V (2023) India's devastating monsoon season is a sign of things to come, as climate and poor planning combine. AP News. https://apnews.com/article/monsoons-climate-change-disasters-india-himalayas-landslides-rains-b1d095811f252ab82268dbefbeef5b3a
Pelckmans I, Belliard J-P, Gourgue O, Dominguez-Granda LE, Temmerman S (2024) Mangroves as nature-based mitigation for ENSO-driven compound flood risks in a large river delta. Hydrol Earth Syst Sci 28:1463–1476. https://doi.org/10.5194/hess-28-1463-2024
Peters A (2017) Moondani Yulenj: an examination of aboriginal culture, identity and education [PDF]. Swinburne University of Technology. https://www.swinburne.edu.au/media/swinburneeduau/library/research-help/referencing/files/APA7-Guide.pdf
Posner S (2021) The rights of nature—can an ecosystem bear legal rights? State of the Planet. https://news.climate.columbia.edu/2021/04/22/rights-of-nature-lawsuits/
Pulkki J, Keto S (2022) Ecosocial autonomy as an educational ideal. Relations Beyond Anthropocentrism 10(2):75–90. https://doi.org/10.7358/rela-2022-02-puke
Rambaree K, Powers MCF, Smith RJ (eds) (2023) Ecosocial work in community practice: embracing ecosocial worldviews and promoting sustainability. Routledge, London
Ramsay S, Boddy J (2017) Environmental social work: a concept analysis. Br J Soc Work 47(1):68–86
RAND Corporation (2024) Adaptation vs. mitigation of climate change: what do developing countries need? RAND Commentary
Richardson K, Steffen W, Rockström J, Cornell SE, Fetzer I, Bennett EM, Biggs R, Carpenter SR, de Vries W, Gerten D, Gleeson EH (2023) Earth beyond six of nine planetary boundaries. *Science*. Advances 9(37):eadh2458. https://doi.org/10.1126/sciadv.adh2458
Rockström J, Steffen W, Noone K, Persson Å, Chapin FS, Lambin E, Lenton TM, Scheffer M, Folke C, Schellnhuber HJ, Nykvist B (2009) Planetary boundaries: exploring the safe operating space for humanity. Ecol Soc 14(2):32. https://www.ecologyandsociety.org/vol14/iss2/art32/
Salonen AO, Konkka J (2015) An ecosocial approach to well-being: a solution to the wicked problems in the era of Anthropocene. Foro de Educación 13(19):19–34
Schaefer J (2023) Ecological ethics. In: Wolfe BN et al (eds) St Andrews encyclopaedia of theology. St Andrews University. Retrieved from https://www.saet.ac.uk/Christianity/EcologicalEthics
Sierra Club v. Morton, 405 U.S. 727 (1972)
Steffen W, Richardson K, Rockström J, Cornell SE, Fetzer I, Bennett EM, Biggs R, Carpenter SR, de Vries W, de Wit CA, Folke C (2015) Planetary boundaries: guiding human development on a changing planet. Science 347(6223):1259855. https://doi.org/10.1126/science.1259855

Stockholm Resilience Centre (2023) New report: tipping point threats and opportunities accelerate. Stockholm Resilience Centre. Retrieved from https://www.stockholmresilience.org
Stone CD (2010) Should trees have standing?: law, morality, and the environment, 3rd ed (first edition 1972) edn. Oxford University Press, Oxford
Thom D, Seidl R (2016) Natural disturbance impacts on ecosystem services and biodiversity in temperate and boreal forests. Biol Rev Camb Philos Soc 91(3):760–781. https://doi.org/10.1111/brv.12193
Tol WA, Song S, Jordans MJD, Reis R (2013) Strengths-based community action as a source of resilience for children affected by armed conflict. Child Adolesc Psychiatr Clin N Am 22(2):365–382. https://pmc.ncbi.nlm.nih.gov/articles/PMC5314737/
Turkelboom F, Demeyer R, Vranken L, De Becker P, Raymaekers F, De Smet L (2021) How does a nature-based solution for flood control compare to a technical solution? Case study evidence from Belgium. Ambio 50(8):1431–1445. https://doi.org/10.1007/s13280-021-01548-4
Tyagi P (2022) Flood risks, mangrove ecosystem and coastal megacities: assessing the adaptation benefits of mangrove ecosystem to urban flood risk in the H/E Ward of Mumbai. Indian J Environ Protect 42(6):643–651
U.S. Environmental Protection Agency (2025) Persistent organic pollutants: a global issue, a global response. https://www.epa.gov/international-cooperation/persistent-organic-pollutants-global-issue-global-response
UNEP (2024) Global resources outlook 2024: bend the trend—pathways to a liveable planet as resource use spikes (IRP report). UNEP. https://www.unep.org/resources/global-resource-outlook-2024
University of Exeter (n.d.) Religion and ecology: Christian attitudes. University of Exeter. Retrieved https://humanities.exeter.ac.uk/theology/research/projects/faithandecology/christianity/
Vaughan A (2016) Human impact has pushed earth into the anthropocene, scientists say. The Guardian. Retrieved from https://www.theguardian.com/environment/2016/jan/07/human-impact-has-pushed-earth-into-the-anthropocene-scientists-say
Vrikki P (2024) Measuring up? The illusion of sustainability and the limits of big tech self-regulation. Sustainability 16(23):10197. https://doi.org/10.3390/su162310197
Waddock S (2024) Holistic eco-social imaginaries for a life-centered future. Sustain Sci 19(6):2119–2134
Wang P, Has E (2021) Local perspective and practical knowledge: social work practice research on chronic diseases of the nomadic area in China. Asian Soc Work Policy Rev 15(1):60–66
Wessells MG (2016) Strengths-based community action as a source of resilience for children affected by armed conflict. Global Mental Health 3:e1. https://doi.org/10.1017/gmh.2015.23
White L (1967) The historical roots of our ecologic crisis. Science 155(3767):1203–1207. https://doi.org/10.1126/science.155.3767.1203
Willkofer F, Wood RR, Ludwig R (2023) Assessing the impact of climate change on high return levels of peak flows in Bavaria applying the CRCM5 large ensemble. EGUsphere 28:2969–2989. https://doi.org/10.5194/egusphere-2023-2019
Xian Z, Nakaya T, Liu K, Zhao B, Zhang J, Zhang J, Lin Y, Zhang J (2024) The effects of neighbourhood green spaces on mental health of disadvantaged groups: a systematic review. Humanit Soc Sci Commun 11(1):488. https://doi.org/10.1057/s41599-024-02970-1

Chapter 3
Exploring the Research Trends and Insights on the Role of Social Work in Disaster Management

T. Adarsh and K. P. Haneena Sulthana

Abstract Disaster management has emerged as a critical area of inquiry within the discipline of social work, reflecting the increasing vulnerability of communities to natural and human-induced hazards. This chapter presents a bibliometric study on the research trends of disaster management in social work, highlighting the evolution and intellectual structure of this emerging interdisciplinary field. The study employed a bibliometric research design, which applies quantitative techniques to examine publication patterns, thematic evolution, and knowledge networks within the domain. Data were retrieved from the leading academic database Scopus, yielding a dataset of 1469 documents across different sources, including peer-reviewed journals, conference proceedings, and book chapters. The analysis employed bibliometric techniques using RStudio, which includes publication volume, authorship patterns, citation metrics, and keyword co-occurrence to examine the structural and thematic development of the field. The findings reveal a steadily expanding body of scholarly work, characterized by interdisciplinary linkages and a gradual shift toward practice-oriented research addressing resilience and community engagement. The annual scientific production shows a steady growth over the period from 1990 to 2024. This upward trajectory highlights the growing recognition of disaster management as a pressing social issue. This chapter contributes to the understanding of how social work scholarship addresses disaster management, offering insights into its past developments, current trajectories, and future research directions.

Keywords Social work · Disaster management · Bibliometric analysis · Thematic shift · Research trends

---

T. Adarsh (✉)
Department of Geography, Sree Sankaracharya University of Sanskrit,
Kalady, Ernakulam, Kerala, India
e-mail: adarshgeo@ssus.ac.in

K. P. Haneena Sulthana
Department of Psychology, Sree Sankaracharya University of Sanskrit,
Kalady, Ernakulam, Kerala, India
e-mail: Haneenasulthana@ssus.ac.in

J. K. Joseph, A. P. Pradeepkumar (eds.), *Eco-Social Work for Disaster Risk Reduction*,
Disaster Resilience and Green Growth, https://doi.org/10.1007/978-981-95-4885-9_3

3.1 Introduction

Social work, fundamentally concerned with enhancing individual and community well-being, has an elusive definition that varies across cultural and political contexts (International Federation of Social Workers 2014). The profession is characterized by a holistic approach that integrates various facets of human experience, including psychological, social, and environmental factors (Ornellas et al. 2016).

Social work plays a crucial role in diverse domains such as community service, humanitarian action, and disaster recovery, characterized by its commitment to justice, welfare, and psychosocial support. The International Federation of Social Workers (IFSW) provides a globally recognized definition that emphasizes the mission of social work to promote social change and development, social cohesion, and the empowerment and liberation of people. This definition indicates the necessity for social workers to operate in a context that takes into account the macro and structural influences on societal functioning (Ornellas et al. 2016). It highlights the importance of academic rigor in social work education, advocating for a framework that is informed by empirical research and theory.

Disaster management plays a critical role in minimizing the adverse impacts of natural and human-made hazards on lives, property, and the environment. By enhancing preparedness, ensuring efficient response, and fostering community resilience, disaster management reduces vulnerability and safeguards livelihoods. It also contributes to sustainable development by embedding risk reduction into policies and practices (United Nations Office for Disaster Risk Reduction 2015). Disaster management emphasizes coordination among stakeholders, governments, NGOs, and communities, making recovery and rehabilitation more effective (Shaluf 2008).

The intersection between social work and disaster management is increasingly recognized as essential for effective crisis intervention and recovery strategies. Social workers play critical roles before, during, and after disasters, encompassing a wide array of responsibilities that are integral to disaster preparedness and response (Drolet et al. 2021).

Firstly, during the preparedness phase, social workers act as educators and community organizers, equipping communities with knowledge about disaster risks and response strategies. They collaborate with various agencies to enhance community resilience and social capital, which are vital for effective disaster management (Machimbidza et al. 2022). The ability of social workers to engage vulnerable populations helps tailor preparation efforts that acknowledge and address unique community needs (Rowlands 2013; Pyles 2007).

During the disaster response phase, social workers provide critical support through crisis counselling and trauma treatment. This involves not only assisting affected individuals but also coordinating multi-agency responses to ensure that services are comprehensive and holistic (Drolet et al. 2021). The provision of mental health and psychosocial support during disasters is essential, and timely interventions significantly improve recovery outcomes, as established in various studies (Naturale 2007; Wong et al. 2021).

Research suggests that the provision of continuous professional supervision and adequate resources is crucial for social workers engaging in post-disaster recovery efforts (Hay et al. 2021). This phase often involves facilitating community rebuilding processes, promoting social development, and fostering resilience through ongoing support systems (Hay et al. 2023; Ladapasea et al. 2021), which are critical as communities seek to regain their sense of normalcy. The development of training curricula focused on disaster recovery for social workers aims to equip them with the necessary skills and theoretical knowledge to address the complexities of working in post-disaster settings (Rowlands 2013; Huang et al. 2013).

Social workers can play an important role in shaping global refugee policies, ensuring that humanitarian efforts align with social justice principles (Harding and Libal 2011). Social workers are equipped with skills that extend beyond immediate crisis intervention. They provide essential psychosocial support during and after disasters. Their training enables them to address not only the visible needs of affected individuals but also the underlying psychological and social dynamics that can significantly hinder recovery efforts (Hay and Pascoe 2022).

Studies indicate that social workers are essential in enhancing disaster risk reduction strategies, advocating for marginalized communities, and ensuring that recovery efforts prioritize inclusivity and equitable access to resources (Beltran et al. 2021; Hussain et al. 2020). Research in various contexts illustrates how social capital, often utilised by social workers, can facilitate community cohesion and collective action during disaster response, leading to more effective recovery outcomes (Machimbidza et al. 2022). Comprehensive training and intentional collaboration among social workers, emergency management professionals, and community organizations are essential for improving the overall efficacy of disaster management strategies (Peek et al. 2020).

The integration of social work in disaster management not only facilitates immediate response efforts but also lays the groundwork for long-term recovery and resilience building within communities. By actively engaging across all phases of disaster management, social workers underscore their vital role in strengthening preparedness, coordinating responses, and promoting sustainable recovery (Rowlands 2013).

As research in this area expands across disciplines, bibliometric analysis provides a systematic approach to mapping publication trends, influential authors, collaborative networks, and thematic developments (Aria and Cuccurullo 2017). Bibliometric analysis has emerged as a crucial tool for understanding research trends, evaluating scientific contributions, and mapping knowledge networks across diverse fields (Passas 2024; Kumar 2025). The applications of bibliometric techniques reveal not only authorship and publication patterns but also citation relationships among documents, offering insights into collaborative frameworks and thematic development within disciplines (Donthu et al. 2021). This study, therefore, examines the evolving landscape of disaster management in social work, highlighting emerging patterns and identifying gaps for future research.

This chapter aims to review existing evidence on the intersection of disaster management and social work. This is timely for several reasons. First, disaster management has become a multidisciplinary field, yet existing reviews often overlook the distinct contributions of social work in crisis intervention and community resilience. Second, this review synthesizes key contributions in this field to highlight how social workers engage in preparedness, response, recovery, and mitigation processes. Third, the study is timely in identifying emerging trends and research gaps, thereby guiding future scholarly and practical efforts to strengthen social work's role in disaster management. Using the Elsevier Analytics Scopus database, the study identifies and analyses key publications, clustering them to reveal thematic patterns and knowledge structures. The remainder of this chapter is structured as follows: The next section outlines the research method (Sect. 3.2). Section 3.3 presents the result and discussion of the bibliometric analyses. Finally, Sect. 3.3.10 concludes the chapter.

3.1.1 Need and Significance

Disasters, both natural and human-made, pose profound challenges to individuals, families, and communities, with social work playing a central role in providing psychosocial support, coordinating relief, and fostering resilience (Dominelli 2012). Despite this, research in disaster management within social work is scattered across diverse domains such as community resilience, trauma-informed care, humanitarian response, and environmental justice. There is thus a critical need to systematically map the knowledge structure and research trends in this interdisciplinary field. The study emphasizes the need for a more comprehensive approach to building resilience and preparedness within communities by identifying gaps, for instance, the limited focus on disaster preparedness within social work education and the uneven integration of mental health and psychosocial support strategies (Ronan & Johnston 2017).

3.1.2 Objectives

- To analyse publication trends in disaster management research within the field of social work
- To identify the most influential contributions in this domain, including leading authors, highly cited publications, key institutions, and contributing countries
- To examine patterns of scholarly collaboration through co-authorship networks at the author, institutional, and country levels
- To map the intellectual and conceptual structure of disaster management in social work using co-citation and co-word analyses, highlighting dominant themes and emerging research areas

3.2 Method

This study employed a bibliometric research design, which uses quantitative techniques to analyse patterns of scientific publication and map the intellectual and conceptual structure of a research field (Donthu et al. 2021). The bibliometric analysis of literature related to this domain reveals a focused yet steadily expanding body of scholarly work addressing the most consequential works in the field of social work. Bibliometric analysis is a robust quantitative method used to assess the structure, performance, and thematic evolution of a scientific field by applying statistical and network-based tools (Adarsh and Haneena Sulthana 2025). A total of 1469 documents were retrieved from distinct sources, including peer-reviewed journal articles, book chapters, conference proceedings, and reviews, capturing the interdisciplinary and evolving nature of research on this topic (Fig. 3.1).

The search was constructed using a dual-topic query to capture the relevant data. The data collection stage involves extracting relevant research articles from selected databases using well-defined search queries. The search query used to extract documents from Scopus is listed below.

TITLE-ABS-KEY = ((“Social Work” OR “Social Worker*” OR “Community Social Work” OR “Environmental Social Work” OR “Disaster Social Work” OR “Humanitarian Social Work” OR “Psychosocial Support” OR “Trauma-Informed Care” OR “Mental Health Social Work” OR “Social Work Practice”) AND (“Disaster” OR “Disaster Risk Reduction” OR “Disaster Management” OR “Disaster Preparedness” OR “Disaster Response” OR “Disaster Recovery” OR “Emergency Management” OR “Crisis Response” OR “Crisis Management” OR “Risk Governance” OR “Resilience Building” OR “Community Resilience”)).

3.2.1 Data Analysis

The analysis was conducted using specialized bibliometric tools, Biblioshiny (an R-based bibliometric package) (Aria and Cuccurullo 2017). The following techniques were applied:

- Descriptive analysis: This involved the identification of publication trends, prominent journals, influential authors, and citation metrics.
- Co-occurrence analysis: Keywords were analysed to visualize the thematic structure of the research field using network maps.
- Thematic mapping: Themes were classified based on relevance and development degree using a two-dimensional analysis, categorizing topics into Basic Themes, Motor Themes, Niche Themes, and Emerging or Declining Themes.
- Thematic evolution analysis: Temporal changes in research focus were mapped using Sankey diagrams to trace the progression of topics over different time periods.

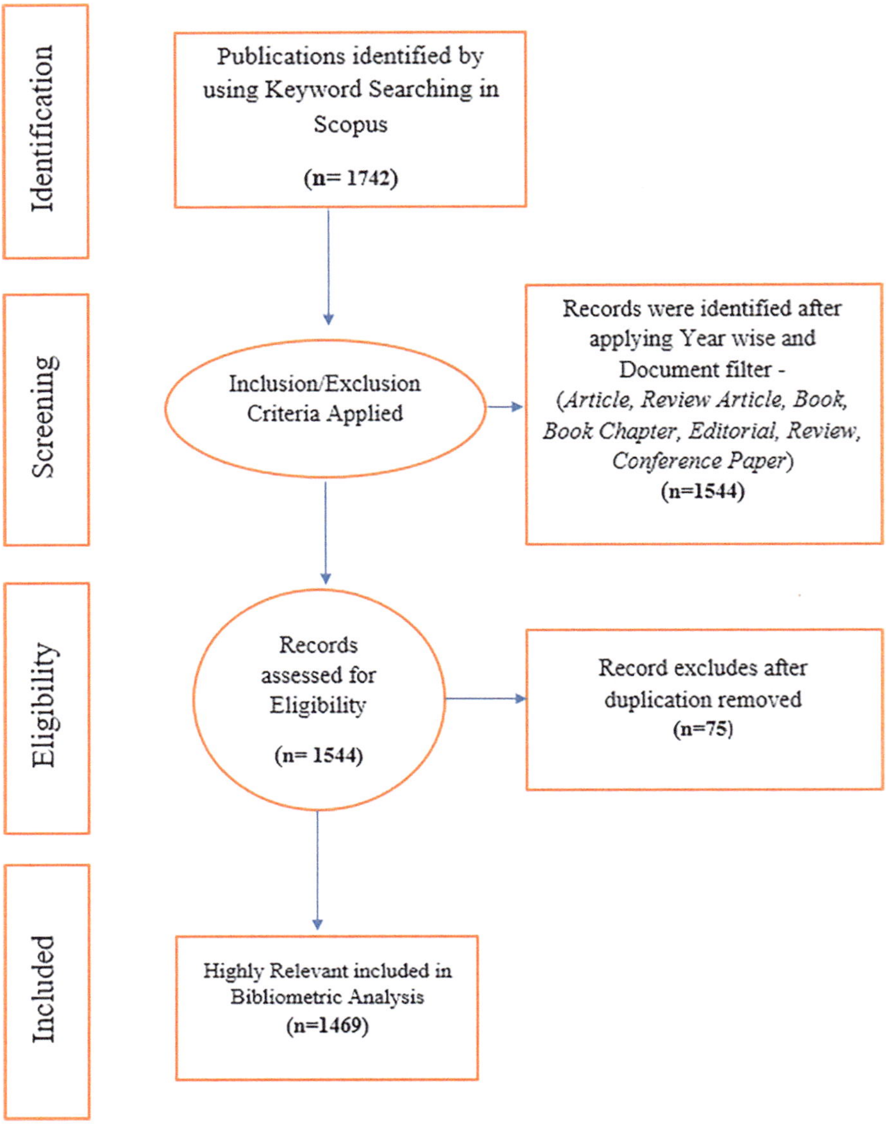

Fig. 3.1 Methodology flow chart—bibliometric research design

- Factorial analysis: Applied to identify underlying structures within the data by performing multiple correspondence analyses, thereby revealing thematic clusters.
- Collaboration network analysis: Country-level collaboration networks were assessed to identify leading research hubs and international partnerships.

3.3 Result and Discussion

3.3.1 Annual Scientific Production

The annual scientific production between *1990 and 2024* reveals a steady upward trend in publications addressing the intersection of social work and disaster management. Early years (1990–1999) were marked by limited scholarly attention, with publications rarely exceeding 10 per year (Fig. 3.2). A significant breakthrough occurred in *1997*, with *15 publications*, followed by another spike in *2002 (18 publications*).

The mid-2000s marked a turning point. Following global disasters such as the *2004 Indian Ocean Tsunami* and *Hurricane Katrina (2005)*, there was a noticeable surge, with *45 publications in 2006*. This momentum continued, with consistent growth through the 2010s. The *Sendai Framework for Disaster Risk Reduction (2015–2030)* coincided with a marked increase, particularly in *2015 (57 publications)* and *2016 (63 publications)*.

The most dramatic rise occurred in the late 2010s and early 2020s. In *2018, 90 publications* were recorded, and by *2021 the number surged to 132*, reflecting heightened global research attention. This peak corresponds with the *COVID-19 pandemic*, which underscored the critical role of social work in health crises, psychosocial support, and community resilience. The trend continued strongly, with *121 publications in both 2022 and 2023*, culminating in a record *186 publications in 2024*.

This trajectory demonstrates the growing academic recognition of social work as a vital discipline in disaster risk reduction, response, and resilience-building. The exponential rise in publications, particularly after 2010, illustrates how scholarship

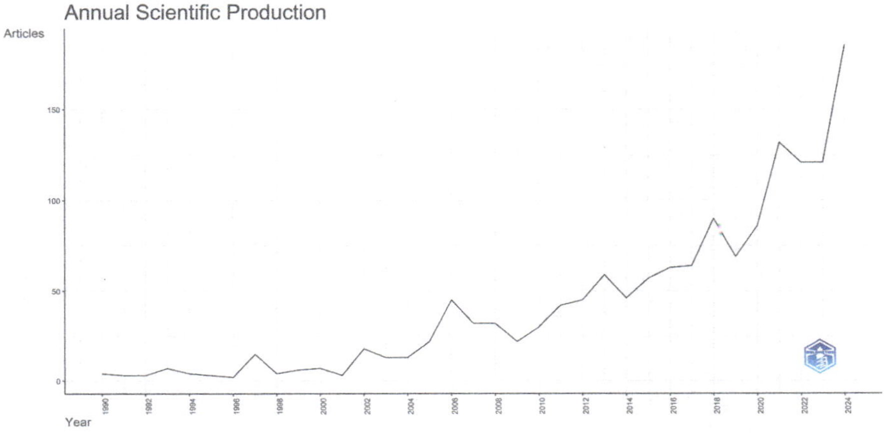

Fig. 3.2 Annual scientific production. (All figures in this chapter, except Fig. 3.6) has been prepared using Biblioshiny (Aria and Cuccurullo 2017)

has become increasingly responsive to global disaster challenges and international policy frameworks.

3.3.2 Average Citations Per Year

The analysis of *average citations per article per year* between *1990 and 2024* demonstrates fluctuating scholarly attention to the theme of *social work practice in disaster contexts*. In the *early 1990s*, there were relatively *few publications*, but some of them received *high average citations*, such as *1992 (67.33 citations per article)* and *1990 (25.75)*, indicating the *foundational impact* of early works.

During the *2000s*, the field saw *moderate but consistent growth*, with notable peaks in *2004 (39.92)*, *2007 (55.78)*, and *2009 (38.41)*, reflecting heightened attention around *global disaster events and evolving frameworks in social work practice*. The *2010s* show a *broadening of research output* with *steady citations*, though the average impact per paper began to stabilize, ranging from *12.81 (2013)* to *24.40 (2011)*. This suggests a *shift from highly cited seminal works to a larger volume of specialized studies*.

In the *post-2018 period*, while the *number of publications sharply increased*, the *average citations per article declined*, for example, in *2021 (12.92)*, *2022 (9.26)*, *2023 (4.92)*, and *2024 (1.36)*. This trend likely reflects the *lag in citation accumulation for recent publications*, alongside the *diversification of research themes* in disaster-related social work. The study indicates that the *1990s and 2000s were citation-rich eras with seminal contributions*, whereas the *2010s onward saw an expansion in publication volume but relatively fewer citations per article*, a typical trend in *maturing research domains*.

3.3.3 Most Relevant Sources

The analysis of source-wise distribution of publications indicates that research on *social work practice and disaster management* has been concentrated in a few leading journals. The *British Journal of Social Work* and *International Social Work* emerged as the most influential sources, each contributing 45 articles to the field. This is followed by *Prehospital and Disaster Medicine*, which published 28 articles, emphasizing the intersection of emergency healthcare and social work in disaster contexts. The *International Journal of Environmental Research and Public Health* accounted for 25 articles, reflecting the growing relevance of public health perspectives in disaster-related social work research.

Another significant platform is the *International Journal of Disaster Risk Reduction*, which has published 20 articles, highlighting the preventive and resilience-building dimensions of social work in disaster settings. Contributions from the *Australian Social Work* journal, with 19 articles, underscore the regional and practice-oriented approaches to disaster interventions. Similarly, the *International Review of Psychiatry* featured 18 articles, pointing to the mental health dimensions closely tied to disaster experiences and social work interventions.

Specialized outlets such as the *Journal of Gerontological Social Work* (16 articles), the *Journal of Social Service Research* (14 articles), and the *Journal of Social Work in Disability and Rehabilitation* (14 articles) have also played important roles by addressing vulnerable groups such as the elderly, people with disabilities, and communities requiring rehabilitative support in disaster contexts.

3.3.4 Core Sources by Bradford's Law

Bradford's Law, introduced by Samuel C. Bradford (1934), *describes the dispersion of articles across scientific journals. The core sources on social work practice in the context of disaster studies between 1990 and 2024 are concentrated in a few highly productive journals*. Among these, the *British Journal of Social Work* and *International Social Work* occupy the top positions, each contributing 45 articles, highlighting their central role in advancing the discourse on disaster-related social work practice (Fig. 3.3).

The *Prehospital and Disaster Medicine* journal stands in third position with 28 articles, reflecting the intersection between medical disaster management and social work interventions. Similarly, the *International Journal of Environmental Research and Public Health* (25 articles) and the *International Journal of Disaster Risk Reduction* (20 articles) emphasize the multidisciplinary approaches that integrate social work with environmental health and disaster risk frameworks.

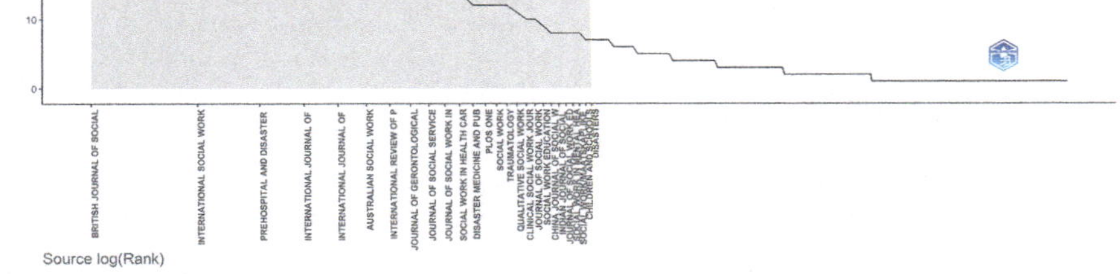

Fig. 3.3 Core sources by Bradford's Law

Australian Social Work (19 articles) and the *International Review of Psychiatry* (18 articles) reveal the significance of regional practice experiences and the mental health dimensions of disaster-affected populations. Specialized contributions are also evident in sources like the *Journal of Gerontological Social Work* (16 articles), *Journal of Social Service Research* (14 articles), and the *Journal of Social Work in Disability and Rehabilitation* (14 articles), which underscore the diverse populations and contexts addressed within the field. *Together, these journals form the core knowledge base for understanding how social work practice has evolved in addressing disaster-related challenges, providing both theoretical insights and practical frameworks.*

3.3.5 Sources' Local Impact

The analysis of sources highlights that the most influential journal in this research domain is *International Social Work*, which holds an h-index of 20, has accumulated 1267 citations, and has published 45 papers since 2005. The *British Journal of Social Work* closely follows with an h-index of 17, 1236 citations, and an equal number of 45 publications, though its contribution to the field dates back to 1990, making it one of the earliest contributors. Another significant contributor is *Australian Social Work*, which has an h-index of 13, 472 citations, and 19 publications beginning from 1991. Similarly, the *International Review of Psychiatry* demonstrates notable influence with an h-index of 13, 445 citations, and 18 publications since 2004.

In recent years, journals such as the *International Journal of Environmental Research and Public Health* have gained prominence, with an h-index of 12, 659 citations, and 25 publications since 2017, reflecting their growing role in disaster-related and health-focused research. Other journals, including the *Journal of Social Service Research*, *Prehospital and Disaster Medicine*, *Social Work in Health Care*, and the *International Journal of Disaster Risk Reduction*, also contribute significantly, with h-indices ranging between 9 and 10, and publication records spanning from the early 1990s to recent years. This distribution indicates that while traditional social work journals remain the backbone of research in this field, interdisciplinary and disaster-focused journals have increasingly gained influence in recent decades.

3.3.6 Most Relevant Authors

The analysis of the most influential authors in the field between 1990 and 2024 highlights a set of key contributors who have shaped scholarship in disaster management, social work, and related interdisciplinary areas. Dominelli L. emerges as the most prolific author, contributing *25 articles* with a significant *fractionalized count of 19.31*, underscoring her leadership in academic discourse. Drolet J.L. follows with *17 articles* (*fractionalized 8.01*), while Alston M.M. has contributed *15 articles* (*fractionalized 9.20*), reflecting their sustained engagement with the subject.

Other prominent contributors include Harms L. (*13 articles, 2.88 fractionalized*) and Sim T. (*10 articles, 4.67 fractionalized*), both of whom have advanced theoretical and applied perspectives in the discipline. Similarly, Wu H. (*10 articles, 5.39 fractionalized*), Boetto H. (*9 articles, 4.12 fractionalized*), and Dückers M.L.A. (*9 articles, 3.33 fractionalized*) have made substantial scholarly inputs, especially in the areas of resilience, social vulnerability, and community-centred responses.

Contributions by Boddy J.H.M. (*8 articles, 2.89 fractionalized*) and Dobalian A. (*8 articles, 1.64 fractionalized*) further strengthen the intellectual foundation of the field, diversifying its methodological and thematic scope. Collectively, these authors form the intellectual core, driving forward research agendas and enhancing knowledge production within the field of social work and disaster studies.

3.3.7 Author Productivity Through Lotka's Law

Lotka's Law, proposed by Alfred J. Lotka (1926), guides the analysis of author productivity, as depicted through Lotka's Law, reveals a highly skewed distribution in scholarly contributions within this research field. The graph demonstrates that the majority of authors have contributed only a single publication, with the proportion of authors sharply declining as the number of written documents increases.

In this case, the curve illustrates that fewer than 10% of authors have written more than three papers, while the overwhelming majority remain single-publication contributors (Fig. 3.4). Such findings highlight the dominance of a small group of prolific authors who shape much of the intellectual discourse, while most

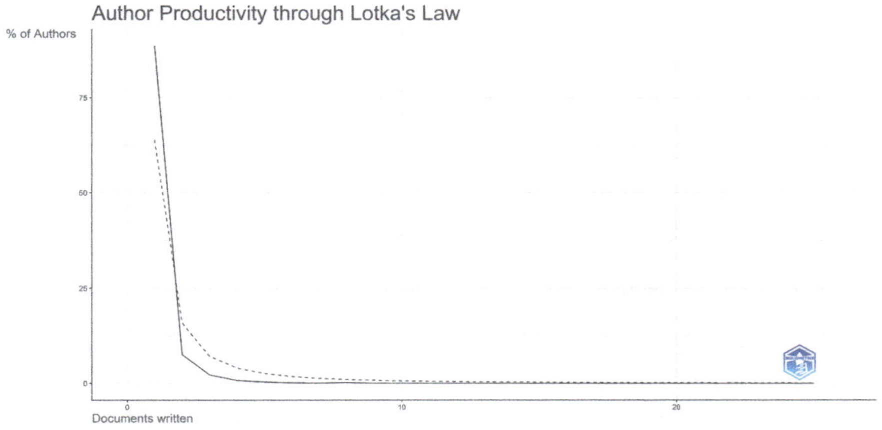

Fig. 3.4 Author productivity through Lotka's Law

researchers participate on a more limited scale. The distribution of author productivity in the study follows a pattern consistent with Lotka's Law. The majority of contributors (3484 authors, accounting for 88.6%) have produced only a single document, reflecting the dominance of occasional contributors in the field.

This distribution has important implications for understanding knowledge production within disaster and social work research. The reliance on a small number of highly active scholars can result in concentrated expertise and continuity in thematic development, but it may also restrict the diversity of perspectives represented in the literature.

3.3.8 Most Relevant Affiliations

The analysis of the most relevant affiliations reveals that research contributions in this domain are highly concentrated among leading global universities. The *University of Melbourne* leads with *27 articles*, followed closely by *Charles Sturt University* and the *University of California*, each contributing *26 articles*. The *Hong Kong Polytechnic University* also shows significant engagement, with *24 articles*, reflecting the institution's growing prominence in this field. Similarly, *Griffith University* makes a notable contribution with *22 articles*.

Other prominent institutions include *Durham University*, *King's College London*, and the *University of Calgary*, each producing *20 articles*, underscoring their active role in advancing scholarship. In addition, *Tulane University* and the *University of Social Welfare and Rehabilitation Sciences* have each contributed *19 articles*, highlighting their commitment to research in disaster management, social work, and related interdisciplinary areas. The study highlights the strong involvement of universities across Australia, North America, Europe, and Asia, reflecting the global nature of research on this theme.

3.3.9 Corresponding Authors' Countries

The distribution of corresponding authors' countries highlights the global nature of research on disaster social work between 1990 and 2024. The *United States* emerged as the most productive country, contributing *26.8%* of the total publications. Interestingly, *99.5%* of these were single-country publications (SCP), indicating a strong tradition of independent research, with only *0.5%* involving international collaboration (MCP) (Fig. 3.5).

Australia followed with *9.5%* of the contributions, demonstrating a notable collaborative research culture. Almost *31.7%* of Australian publications were multi-country publications, showing the country's active engagement in international partnerships. *Canada* contributed *6.4%* of articles, with the highest proportion of international collaboration among the leading countries. Nearly *37.2%* of Canadian

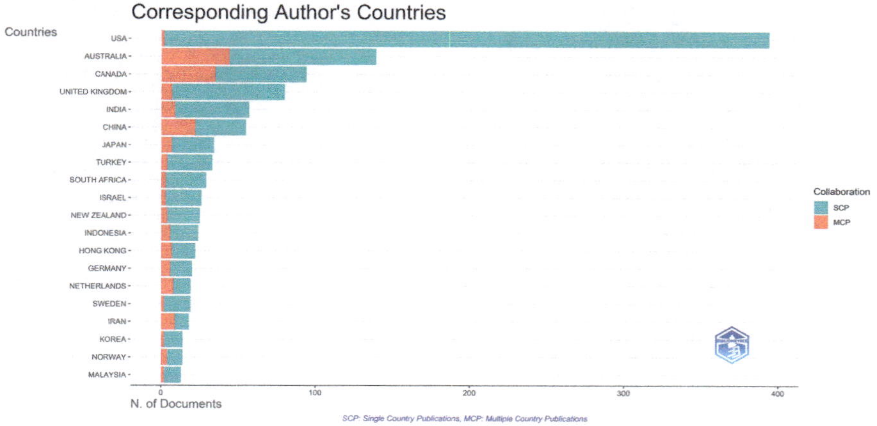

Fig. 3.5 Corresponding authors' countries. The USA is in the top and India is in the fifth position

research output came through multi-country publications, highlighting Canada's significant role in fostering cross-border knowledge exchange in disaster social work.

The *United Kingdom* ranked next with *5.4%* of publications. Despite its moderate output, the UK displayed a strong inclination toward single-country publications (*91.2%*), with limited multi-country collaboration (*8.8%*). *India*, accounting for *3.9%* of the contributions, displayed a balance between domestic research (*84.2% SCP*) and international cooperation (*15.8% MCP*), reflecting its growing presence in global disaster social work research. Similarly, *China* contributed *3.7%* of the publications, with a high share of international collaboration (*40% MCP*), indicating its increasing integration into the global research network.

Other notable contributors included *Japan (2.3%)*, with moderate collaboration (*20.6% MCP*), *Turkey (2.2%)*, *South Africa (2%)*, and *Israel (1.8%)*, all of which reflected relatively smaller but important roles in advancing disaster social work research. *Overall, the findings suggest that while the USA dominates in terms of research volume, countries like Canada, Australia, and China play a vital role in driving international collaboration within this field.*

3.3.10 Countries' Scientific Production

The analysis of scientific production across countries reveals the USA as the dominant contributor with a frequency of 1250 publications, highlighting its leading role in this field. Australia follows with 293 publications, reflecting its strong engagement in research, while the United Kingdom also makes a significant contribution with 212 publications.

Canada accounts for 197 publications, further emphasizing the strong research output from North America. In Asia, China emerges as a major contributor with 139 publications, closely followed by India with 128 publications. Japan also plays an important role with 105 publications.

European contributions are notable, with the Netherlands producing 77 publications. Turkey contributes 75 publications, and Iran accounts for 65 publications, underscoring the growing research involvement of these regions. The distribution indicates a clear dominance by developed nations, while emerging countries like India, China, Turkey, and Iran are steadily strengthening their global research presence.

3.3.11 Most Cited Countries

An assessment of citation impact reveals interesting global patterns. The USA leads with a total of 7976 citations, averaging *20.20 citations per article* (Fig. 3.6). This underscores its consistent research visibility and global influence. Australia follows with 1954 citations and an average of *14.10 citations per article*, while the United Kingdom secures a strong position with 1561 citations and an impressive *19.50 citations per article*. Canada contributes 1140 citations (*12.10 average*), and China demonstrates both scale and impact with 1066 citations, achieving an average of *19.40 citations per article*.

India shows a growing presence with 967 citations and an average of *17.00 citations per article*. Interestingly, Georgia stands out with 739 citations but an exceptionally high *average of 123.20 citations per article*, indicating the remarkable influence of a comparatively smaller volume of research. Other notable contributors include Switzerland, with 460 citations and a high average of *38.30 citations per*

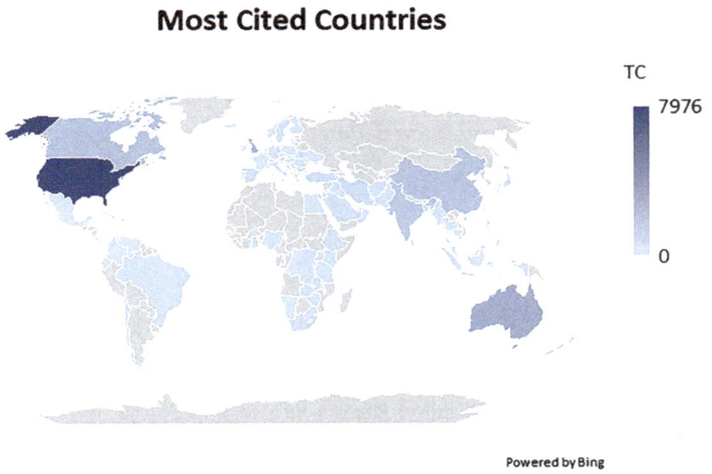

Fig. 3.6 Most cited countries

article, and Israel, with 406 citations (*15.60 average*). The Netherlands also maintains significant visibility, recording 377 citations with an average of *19.80 citations per article*.

While established research hubs like the USA, UK, and China dominate in volume, countries such as Georgia and Switzerland distinguish themselves with *extraordinary per-article impact*, highlighting the quality-driven influence of their contributions.

3.3.12 Most Global Cited Documents

The analysis of the most highly cited documents provides insights into the intellectual foundations and thematic priorities of the field. The top ten most cited works span across social work, psychology, medicine, and public health, indicating the interdisciplinary nature of disaster-related and trauma-informed research.

At the forefront, Bride (2007) published in *Social Work* leads the global citation ranking with *630 citations*. This seminal work emphasizes secondary traumatic stress among social workers, establishing a conceptual and methodological foundation for subsequent studies in disaster response and trauma care. Closely following, Hawkins (2010) in the *British Journal of Social Work* received *497 citations*. This article explores resilience and recovery in the aftermath of Hurricane Katrina, highlighting the intersection of disaster, poverty, and community vulnerability.

From the medical sciences, Evans (2020) in the *New England Journal of Medicine* stands out with *342 citations*. As a high-impact piece written during the COVID-19 pandemic, it underscores the critical role of domestic violence interventions amidst global crises, reflecting the pandemic's broad societal implications. Other influential contributions include Marsac et al. (2016) in *JAMA Pediatrics* with *242 citations*, which focuses on paediatric posttraumatic stress and healthcare system responses, and Mathbor (2007) in *International Social Work* with *237 citations*, addressing community preparedness and resilience in disaster contexts.

Boscarino (2004) (*International Journal of Emergency Mental Health*) and Van Ommeren (2005) (*Bulletin of the World Health Organization*), with *237* and *228 citations*, respectively, demonstrate the importance of integrating mental health care into emergency and humanitarian settings. Contributions from the environmental and psychological sciences also enrich the field. For instance, Tang (2017) in the *International Journal of Environmental Research and Public Health* with *210 citations* reflects growing awareness of the psychosocial dimensions of natural disasters, while Joseph (1992) in the *British Journal of Clinical Psychology* with *199 citations* remains a foundational work in trauma studies. At the last Adams (2008) in *Research on Social Work Practice* with *163 citations* emphasizes organizational and practice-level responses to traumatic stress in social work.

Fig. 3.7 Word Cloud

3.3.13 Most Frequent Words

The analysis of keyword frequency reveals the dominant themes in the research corpus. The most recurring word is "*human*" with 639 occurrences, followed by "*humans*" (465 occurrences) and "*article*" (412 occurrences). The prominence of "*social work*" (402 occurrences) underscores the critical role of social work in disaster-related research (Fig. 3.7).

The frequent appearance of "*disaster*" (387 occurrences) and "*disasters*" (234 occurrences) highlights the strong thematic orientation of the literature toward disaster studies. In terms of demographic focus, "*female*" (250 occurrences), "*male*" (233 occurrences), and "*adult*" (210 occurrences) appear frequently, reflecting an emphasis on gender and age dimensions in research. Additionally, "*mental health*" (211 occurrences) emerges as a recurring keyword, suggesting a growing recognition of psychological and social consequences in disaster contexts.

The frequent words indicate that the literature predominantly engages with themes of human vulnerability, social work, disasters, and mental health, often analysed through gender and demographic perspectives.

3.3.14 Trend Topics

The trend topics analysis highlights how research focus on the intersection of social work and disaster management has evolved over time (1990–2024). The early studies (1990s to early 2000s) concentrated on foundational areas such as "*survival*," "*occupational diseases*," and "*middle age*." These reflect the initial stages of scholarship, where disaster-related research was still exploratory and largely descriptive.

From the mid-2000s onward, the focus shifted to mental health and psychosocial dimensions, with key topics including "*posttraumatic stress disorder*," "*stress disorders*," "*psychological adaptation*," and "*social support*." This indicates an increasing recognition of the psychosocial consequences of disasters and the role of social work in providing trauma-informed care. Between 2010 and 2016, topics such as "*disaster planning*," "*emergency health services*," "*crisis intervention*," and "*terrorism*" gained visibility, reflecting global concerns about preparedness and the integration of social work into emergency response frameworks.

In recent years (2017–2024), the emphasis has expanded toward public health crises and pandemics, with frequent terms such as "*pandemic*," "*COVID-19*," "*COVID-19 pandemic*," and "*family violence*." These emerging keywords demonstrate how the COVID-19 era reshaped research priorities, particularly highlighting vulnerabilities related to mental health, domestic violence, and community resilience during prolonged crises. The trajectory reveals a clear evolution from survival and health impacts toward psychosocial care, resilience, and pandemic-related challenges, positioning social work as a central discipline in disaster risk reduction and recovery.

3.3.15 Co-occurrence Network

The co-occurrence network of keywords provides significant insights into the intellectual structure and thematic clusters within the research domain. The visualization shows three major communities of keywords (Fig. 3.8), each representing distinct but interconnected thematic foci. The largest red cluster revolves around terms such as *human*, *humans*, *article*, *male*, *female*, and *mental health*, suggesting a

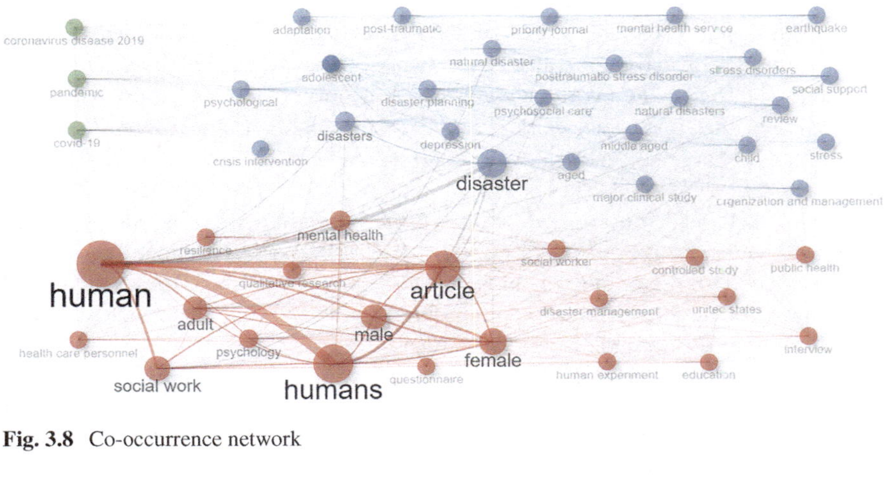

Fig. 3.8 Co-occurrence network

concentration of research on human-centered and psychosocial dimensions, including gender, psychology, social work, and resilience.

The second cluster, highlighted in blue, is organized around the keyword *disaster* and includes terms such as *natural disaster*, *disaster planning*, *posttraumatic stress disorder*, *social support*, and *mental health services*. This cluster reflects the scholarly emphasis on disaster management, psychosocial care, and the impact of crises on mental health, with a strong connection to stress-related outcomes and health services. A smaller but important green cluster focuses on *pandemic*, *COVID-19*, and *coronavirus disease 2019*, capturing the recent surge of literature addressing global health emergencies and their societal consequences.

The network also reveals the strong interconnections across clusters, with keywords such as *mental health* and *disaster* acting as bridges that link human-centered research with broader disaster and pandemic contexts. The varying sizes of the nodes reflect the frequency of keyword use, with *human*, *humans*, and *disaster* being the most dominant, indicating their central role in the discourse. The visualization underscores that while disaster and pandemic-related research form critical thematic axes, human and psychosocial dimensions remain at the core of the field, highlighting an interdisciplinary integration of health, psychology, and disaster management scholarship.

3.3.16 Thematic Map

The thematic map provides a structured visualization of research themes within the field, categorized into four quadrants based on their *development degree (density)* and *relevance degree (centrality)*. In the Motor Themes quadrant, highly relevant and well-developed topics dominate the discourse. These include "human," "humans," "female," "male," and "COVID-19/pandemic," which reflect central and mature areas of scholarship (Fig. 3.9), particularly in connection to disaster management, healthcare, and social dynamics. Closely aligned with these are themes such as "hospital," "patient care," "disaster management," and "climate change," which represent strongly integrated and actively evolving research directions.

The Basic Themes quadrant encompasses concepts of *high centrality but lower development*, such as "social work practice," "sustainable development," "disaster relief," and "emergency services." These topics are essential for the broader field, serving as foundational areas of study, yet require further scholarly depth and integration to reach the maturity of motor themes. In the Niche Themes quadrant, highly developed but less central themes emerge, including "vicarious trauma," "secondary trauma," "social policy," and "displacement." While specialized, these areas hold importance for targeted research communities, suggesting a strong but localized influence rather than broad disciplinary integration.

The Emerging or Declining Themes quadrant contains underdeveloped and less central topics, such as "posttraumatic stress," "disaster social work," "social work students," and "disaster risk reduction." These themes may either represent

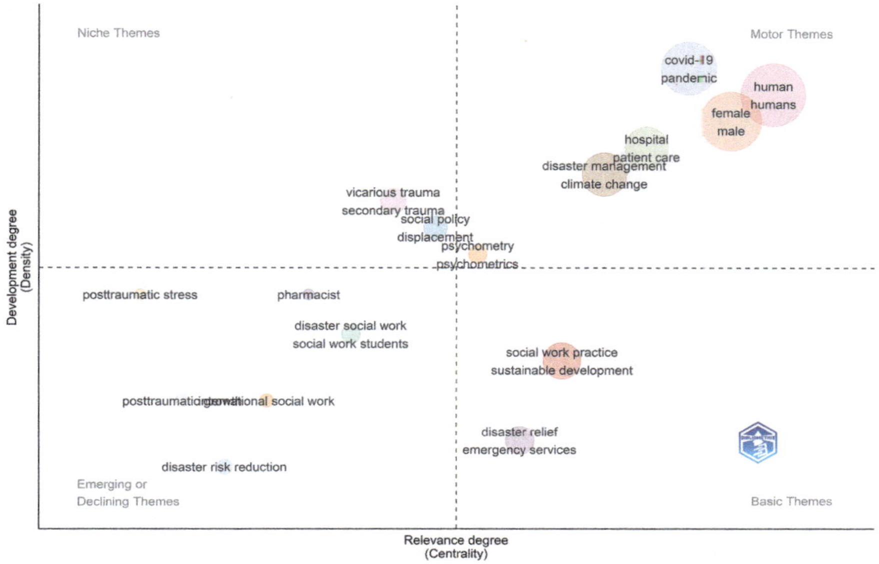

Fig. 3.9 Thematic map

declining areas of interest or nascent fields that could evolve with further scholarly attention. The thematic structure reveals that the field is strongly shaped by *human-centered and pandemic-related themes*, while also maintaining critical connections to disaster management, healthcare, and sustainable development. At the same time, opportunities for growth remain in underdeveloped and emerging areas such as *disaster risk reduction* and *specialized aspects of social work research*.

3.3.17 Thematic Evolution

The strategic diagram of thematic evolution illustrates how the intellectual structure of the field has developed across periods: *1990–1999*, *2000–2009*, *2010–2019*, and *2020–2024*. In the earliest phase (*1990–1999*), research was primarily shaped by fundamental terms such as "human," "major clinical study," "information system," and "review." These keywords reflect the foundational orientation of scholarship, where methodological and clinical considerations were central (Fig. 3.10).

During the *2000–2009* period, the field expanded toward themes associated with disasters and health services. Prominent keywords included "human," "natural disasters," "emergency health service," "PTSD," and "social capital." This phase demonstrates a strong emphasis on linking human-centered concerns with disaster response and psychological outcomes, such as trauma and coping. The persistence

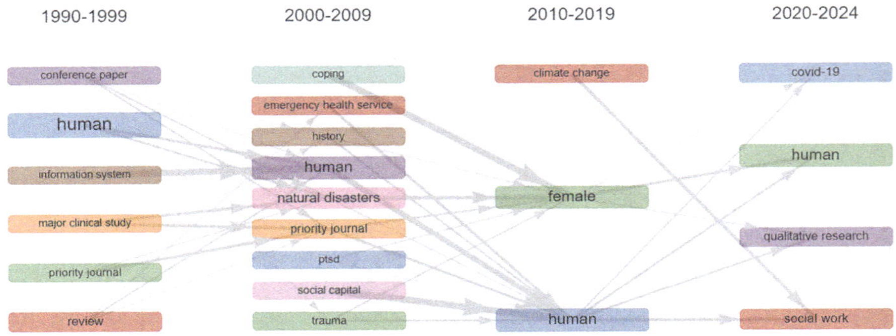

Fig. 3.10 Thematic evolution

of "human" from the earlier decade underscores its role as a central and enduring theme.

In the subsequent period (*2010–2019*), research directions shifted toward broader global challenges, with "climate change" emerging as a significant topic alongside continuing interest in "human" and "female." The introduction of gendered perspectives indicates a diversification of scholarly focus, addressing social vulnerability and differentiated impacts of disasters across populations.

In the most recent phase (*2020–2024*), the influence of the global pandemic is clearly visible, with "COVID-19" emerging as a dominant theme. Alongside this, "human" remains central, while new emphases on "qualitative research" and "social work" highlight methodological diversification and a stronger engagement with applied, practice-oriented dimensions of disaster and health studies.

The thematic evolution demonstrates a clear trajectory from *clinical and methodological foundations (1990s)* to *disaster-centered and psychosocial research (2000s)*, then toward *global challenges and social dimensions (2010s)*, culminating in a strong *pandemic-driven and applied focus (2020s)*. The enduring centrality of "human" across all decades reflects the consistent orientation of the field toward human-centered concerns, while newer themes such as "climate change" and "COVID-19" underscore its adaptability to global crises.

3.3.18 Factorial Analysis

The conceptual structure map generated using the *Multiple Correspondence Analysis (MCA)* method provides a clear visualization of how the intellectual landscape of the field is organized around major thematic clusters. The horizontal axis (*Dim 1*), which explains 78.63% of variance (Fig. 3.11), primarily separates themes related to *health, social work, and disaster management* from those linked to *pandemics, trauma, and resilience*. The vertical axis (*Dim 2*), accounting for 9.40% of

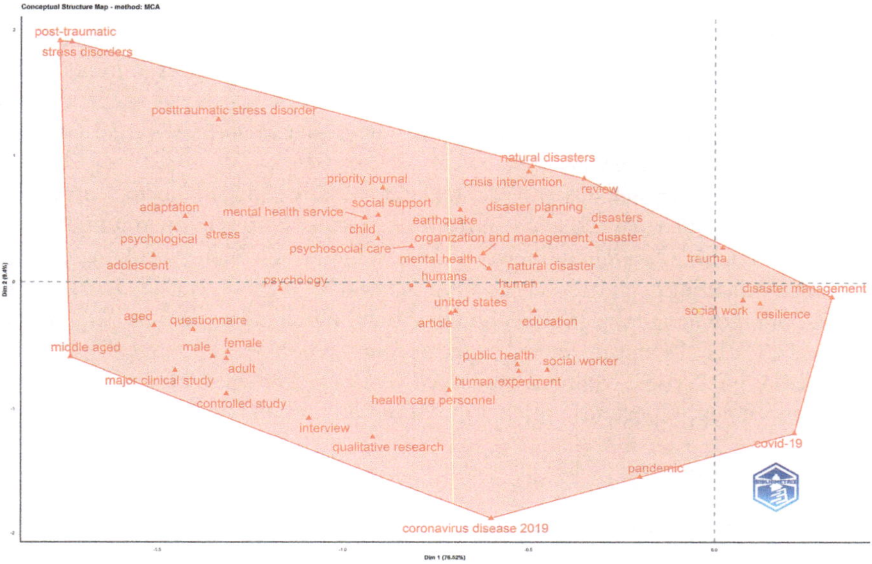

Fig. 3.11 Factorial analysis

variance, distinguishes psychological and clinical concerns from broader social and organizational dimensions.

On the left side of the map, clusters emerge around psychological and clinical themes, including *posttraumatic stress disorders*, *adaptation*, *stress*, *adolescent populations*, *mental health services*, and *psychology*. These terms highlight the strong grounding of the field in psychological well-being, trauma response, and mental health interventions. Nearby, a concentration of demographic and methodological terms such as *male*, *female*, *adult*, *middle-aged*, *questionnaire*, *interview*, and *controlled study* reflects the methodological rigor and the use of diverse population groups in research.

At the centre, highly frequent and general terms such as human, humans, article, education, public health, and social worker form the intellectual core of the field, bridging both the clinical/psychological and the disaster/social domains. This centrality demonstrates the enduring importance of human-focused and interdisciplinary approaches in disaster and health-related scholarship.

On the right side of the map, themes such as disaster management, social work, resilience, trauma, and COVID-19 form a strong cluster that emphasizes contemporary challenges and applied responses. The presence of the pandemic and coronavirus disease 2019 highlights the recent shift in research priorities toward global health emergencies. These terms are closely linked with practice-oriented concerns, such as *disaster planning*, *organization and management*, and *crisis intervention*, showing the integration of theory with practical disaster response and resilience-building strategies.

At the top, terms like posttraumatic stress disorder, priority journal, and social support occupy a specialized position, indicating a niche but influential focus on psychological consequences and support systems in disaster contexts. Meanwhile, methodological advances such as qualitative research and *interviews* appear toward the lower-left quadrant, reflecting a growing interest in interpretive and context-driven research approaches.

The MCA conceptual structure demonstrates a threefold intellectual organization of the field: (1) a *psychological and clinical cluster* emphasizing trauma, stress, and mental health; (2) a *human-centered and methodological core* bridging diverse research orientations; and (3) an *applied disaster and pandemic-focused cluster* highlighting contemporary global challenges and practical responses. The prominence of COVID-19 and resilience suggests that the field has dynamically evolved to address urgent crises, while the continued centrality of human/humans reinforces the consistent anthropocentric orientation of research in this domain.

3.3.19 Collaboration Network

The Country Collaboration Network map illustrates the global research partnerships in this domain, highlighting central hubs and peripheral contributors. The size of the nodes corresponds to the research output and collaborative influence of each country, while the thickness of the connecting lines indicates the strength of bilateral or multilateral collaborations (Fig. 3.12).

At the core of the network, the USA emerges as the most dominant hub, with the largest node size and the densest set of connections. Its strong ties with the United Kingdom, Canada, Australia, India, China, and several European countries position it as the main driver of international collaboration. This reflects the USA's leading

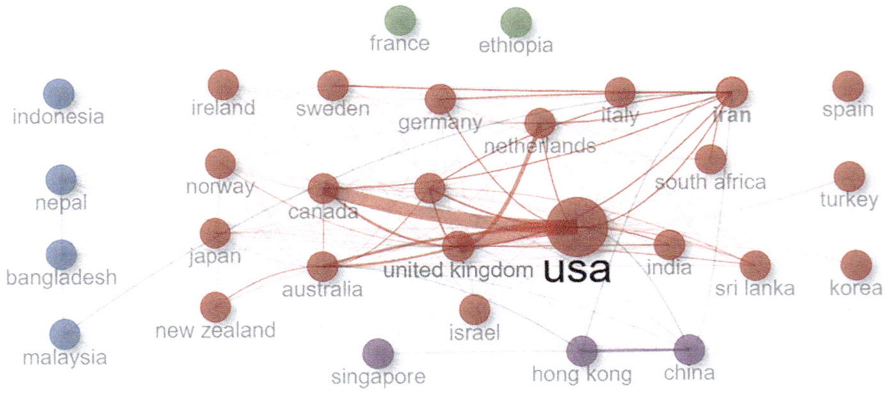

Fig. 3.12 Collaboration network

role in shaping global research agendas, mobilizing resources, and serving as a primary partner for both developed and developing nations.

The United Kingdom, Canada, and Australia form secondary hubs that act as important bridges, facilitating regional and intercontinental collaborations. Their extensive connections with both Western and Asian countries highlight their intermediary role in disseminating knowledge across different regions.

In Europe, countries such as Germany, the Netherlands, Italy, Spain, and Sweden display strong interconnectivity, forming a dense collaborative cluster with North America. This underlines the established transatlantic academic partnerships that continue to dominate global research networks. Similarly, Iran, India, and South Africa stand out as active contributors from Asia and Africa, demonstrating the increasing integration of emerging research nations into global collaborations.

On the periphery, countries such as Nepal, Bangladesh, Malaysia, Indonesia, and Sri Lanka appear less central but nonetheless connected, often through partnerships with larger hubs like the USA, UK, and China. These links suggest a dependency on collaborations with leading countries to gain visibility and contribute to the global discourse.

Smaller clusters, including China, Hong Kong, Singapore, and Israel, form strong intra-regional partnerships while maintaining outward collaborations with larger hubs. This dual strategy highlights both regional strengthening and global integration. Interestingly, France and Ethiopia appear in separate, less connected clusters, suggesting more isolated contributions rather than integration into the broader collaboration web.

The network reveals a hierarchical but interconnected global structure, where the USA leads as the central hub, supported by the UK, Canada, and Australia as secondary anchors, while Europe, Asia, and Africa contribute through varying degrees of integration. Peripheral countries, though less central, play important roles in diversifying perspectives and extending research relevance to local contexts.

3.4 Conclusion

This bibliometric study demonstrates that research at the intersection of social work and disaster management has grown from a fragmented body of work in the 1990s into a robust and rapidly expanding field in the 2010s and 2020s. The exponential rise in publications, particularly in the wake of global disasters and the COVID-19 pandemic, indicates the increasing recognition of social work as a central discipline in disaster risk reduction, resilience-building, and psychosocial care.

The analysis reveals distinct geographical and thematic patterns. While the United States dominates in research volume, countries such as Canada, Australia, and China have emerged as key contributors to international collaboration, thereby enriching the field's global relevance. The evolution of research themes from early descriptive studies of survival and health impacts to complex explorations of trauma, resilience, preparedness, and pandemic response reflects the adaptability of scholarship to shifting global challenges. Co-occurrence, thematic, and conceptual analyses further highlight the interdisciplinary integration of psychology, health, and disaster management, with human-centered concerns and resilience consistently bridging diverse strands of inquiry.

The disaster social work research is characterized by both continuity and dynamism. Continuity in its enduring human-centered orientation, and dynamism in its responsiveness to global crises, methodological diversification, and applied practice needs. Moving forward, greater attention to emerging and underdeveloped themes such as disaster risk reduction, displacement, and sustainable development, alongside stronger international collaboration, will be essential to consolidating the field's contribution to both scholarship and practice in building resilient communities.

References

Adarsh T, Haneena Sulthana KP (2025) Trends and insights in nature-based solutions: bibliometric study. International Journal of Disaster Studies and Climate Resilience 1(1):22–34. https://doi.org/10.64866/j.ijdscr.2025.10003

Aria M, Cuccurullo C (2017) Bibliometrix: an R-tool for comprehensive science mapping analysis. J Informet 11(4):959–975. https://doi.org/10.1016/j.joi.2017.08.007

Beltran S, Luigi P, Kusmaul N, Leon M (2021) Rising above the flood: a systematic review of gerontological social work in disaster preparedness and response. J Gerontol Soc Work 65(5):545–561. https://doi.org/10.1080/01634372.2021.1986764

Bradford SC (1934) Sources of information on specific subjects. Engineering 137:85–86

Dominelli L (2012) Green social work: from environmental crises to environmental justice. Polity Press, Cambridge

Donthu N, Kumar S, Mukherjee D, Pandey N, Lim WM (2021) How to conduct a bibliometric analysis: an overview and guidelines. J Bus Res 133:285–296. https://doi.org/10.1016/j.jbusres.2021.04.070

Drolet J, Lewin B, Pinches A (2021) Social work practitioners and human service professionals in the 2016 Alberta (Canada) wildfires: roles and contributions. Br J Soc Work 51(5):1663–1679. https://doi.org/10.1093/bjsw/bcab141

Harding S, Libal K (2011) Iraqi refugees and the humanitarian costs of the Iraq war: what role for social work? Int J Soc Welf 21(1):94–104. https://doi.org/10.1111/j.1468-2397.2011.00780.x

Hay K, Pascoe K (2022) Engaging social workers in disaster management: case studies from New Zealand. Int J Disaster Risk Reduction 74:102941. https://doi.org/10.1016/j.ijdrr.2022.102941

Hay K, Pascoe K, McCafferty L (2021) Social worker experiences in disaster management: case studies from aotearoa New Zealand. Aotearoa New Zealand Soc Work 33(1):17–28. https://doi.org/10.11157/anzswj-vol33iss1id820

Hay K, Pascoe K, Henley L, Knight F, Stewart K, Radak G (2023) Social work disaster practice. Aotearoa New Zealand Soc Work 35(1):60–70. https://doi.org/10.11157/anzswj-vol35iss1id1011

Huang Y, Fu Y, Wong H (2013) Challenges of social workers' involvement in the recovery of the 5.12 Wenchuan earthquake in China. Int J Soc Welf 23(2):139–149. https://doi.org/10.1111/ijsw.12029

Hussain S, Ibrar M, Alam H (2020) Social work profession and disaster risk reduction in Pakistan. Pakistan J Appl Soc Sci 7(1):29–41. https://doi.org/10.46568/pjass.v7i0.317

International Federation of Social Workers (2014) Global definition of social work. https://www.ifsw.org/what-is-social-work/global-definition-of-social-work/

Kumar R (2025) Bibliometric analysis: comprehensive insights into tools, techniques, applications, and solutions for research excellence. *SEMS* 3:45–62. https://doi.org/10.31181/sems31202535k

Ladapasea E, Anggrainia S, Yantoa A, Baraa Y, Witaka Y, Welina Y et al (2021) Community empowerment psychosocial support for disaster survival children in adonara. JPPMI 1(5):170–178. https://doi.org/10.59247/jppmi.v1i5.22

Lotka AJ (1926) The frequency distribution of scientific productivity. J Wash Acad Sci 16(12):317–323

Machimbidza D, Nyahunda L, Makhubele J (2022) The importance of social work roles in disaster risk management in Zimbabwe. Technium Soc Sci J 27:717–726. https://doi.org/10.47577/tssj.v27i1.5554

Naturale A (2007) Secondary traumatic stress in social workers responding to disasters: reports from the field. Clin Soc Work J 35(3):173–181. https://doi.org/10.1007/s10615-007-0089-1

Ornellas A, Spolander G, Engelbrecht L (2016) The global social work definition: ontology, implications and challenges. J Soc Work 18(2):222–240. https://doi.org/10.1177/1468017316654606

Passas I (2024) Bibliometric analysis: The main steps. *Encyclopedia* 4(2):1014–1025. https://doi.org/10.3390/encyclopedia4020065

Peek L, Champeau H, Austin J, Mathews M, Wu H (2020) What methods do social scientists use to study disasters? An analysis of the social science extreme events research network. Am Behav Sci 64(8):1066–1094. https://doi.org/10.1177/0002764220938105

Pyles L (2007) Community organizing for post-disaster social development. Int Soc Work 50(3):321–333. https://doi.org/10.1177/0020872807076044

Ronan KR, Johnston DM (2017) Promoting community resilience in disasters: the role for schools, youth, and families. Springer, New York. https://doi.org/10.1007/b102725

Rowlands A (2013) Social work training curriculum in disaster management. J Soc Work Disabil Rehabil 12(1–2):130–144. https://doi.org/10.1080/1536710x.2013.78460

Shaluf IM (2008) Disaster types. Disaster Prevention Manag 17(5):704–717

United Nations Office for Disaster Risk Reduction (2015) Sendai framework for disaster risk reduction 2015–2030. United Nations. https://www.undrr.org/publication/sendai-framework-disaster-risk-reduction-2015-2030

Wong L, Alias H, Danaee M, Lee H, Tan K, Tok P et al (2021) Assessment of impact of containment during the covid-19 epidemic and coping behaviours using newly developed assessment tools. Front Public Health 9:787672. https://doi.org/10.3389/fpubh.2021.787672

Chapter 4
Eco-Social Resilience and Nature-Based Solutions: Local Strategies for Disaster Risk Reduction Across the Globe

Megha N Nair, Shibu K. Mani, Irish M. Polinga, and K. R. Baiju

Abstract The gradual transition from traditional grey infrastructure in disaster risk reduction (DRR) has paved the way for the global flourishing of Nature-based Solutions (NbS). It has emerged as a cohesive framework that includes approaches such as ecosystem-based adaptation (EbA), ecosystem-based disaster risk reduction (Eco-DRR), and green/blue infrastructure. These approaches deliver multiple benefits across ecological, social, cultural, and economic sectors while reducing hazard impacts and building eco-social resilience. This chapter explores what NbS means in relation to similar concepts, examines how stakeholder engagement determines project outcomes, and presents diverse global strategies that demonstrate multiple co-benefits. Drawing on global case studies—from urban regeneration initiatives in Europe and Asia to mangrove restoration in Southeast Asia and indigenous knowledge systems in Africa and Japan—this paper shows how NbS can blend traditional practices with modern science to deliver multiple resilience benefits, including poverty alleviation, food security, and climate mitigation. Unlike conventional approaches that focus narrowly on single outcomes, NbS offers integrated gains, making them an essential component of the eco-social dimension of disaster risk reduction (DRR).

Success depends critically on stakeholder perceptions, community participation, and governance structures. While positive perceptions often link to aesthetic and recreational benefits, challenges persist, including limited awareness of co-benefits,

M. N. Nair (✉) · K. R. Baiju
School of Environmental Sciences, Mahatma Gandhi University, Kottayam, Kerala, India
e-mail: mekhannair@mgu.ac.in; baijukr@mgu.ac.in

S. K. Mani
Department of Civil Engineering, School of Engineering and Technology, CHRIST (Deemed to be University), Bengaluru, Karnataka, India
e-mail: shibuk.mani@christuniversity.in

I. M. Polinga
College of Engineering, Architecture and Technology, De La Salle University, Dasmariñas, Philippines
e-mail: impolinga@dlsud.edu.ph

J. K. Joseph, A. P. Pradeepkumar (eds.), *Eco-Social Work for Disaster Risk Reduction*, Disaster Resilience and Green Growth, https://doi.org/10.1007/978-981-95-4885-9_4

difficulties in economic valuation, institutional barriers, and concerns about potential disservices like pests or blocked views. Gender, age, past experiences, and local power dynamics are factors that shape community responses. Economic mechanisms such as payments for ecosystem services in Indonesia and Madagascar, along with continuous monitoring and transparent communication, can strengthen implementation. Without addressing these practical challenges, NbS initiatives risk losing credibility, potentially causing communities to revert to grey infrastructure. As climate risks intensify, NbS offer equitable pathways for building resilience across both Global North and South contexts, making them essential tools for sustainable disaster risk reduction.

Keywords Eco-social resilience · Eco-DRR · Nature-based solutions · Stakeholder engagement · Climate adaptation

Learning Objectives

- To understand Nature-based Solutions (NbS) and related ecosystem-based approaches
- To examine how stakeholder engagement determines NbS outcomes
- To learn from diverse global NbS strategies delivering multiple co-benefits

4.1 Introduction

Over the past few decades, disaster risk reduction (DRR) has undergone a significant transition from reactive responses to proactive strategies. Early efforts primarily focused on structural investments, particularly engineered infrastructure (Murti and Mathez-Stiefel 2019). However, the critical role played by ecosystems in rendering protective and adaptive services was ignored for a long time (Renaud et al. 2013). Since the tsunami of 2004 in the Indian Ocean, however, there has been increasing momentum for mainstreaming ecosystem services into DRR policy through the use of Nature-based Solutions (NbS) (Ghaleh and Ghaleh 2020). From 2010 to 2024, research on NbS experienced rapid expansion with greater international recognition of its value as a sustainable climate adaptation strategy (Adarsh and Haneena Sulthana 2025).

NbS is an umbrella concept that comprises six related neighbouring concepts—ecological engineering/catchment systems engineering, green/blue infrastructure, ecosystem approach, ecosystem-based adaptation/mitigation, ecosystem service approach/framework, and natural capital. Eco-DRR and ecosystem-based adaptation (EbA) are various ecosystem-based approaches for reducing various risks, falling under the concept of NbS (De Silva et al. 2022). Likewise, this concept is designed to effectively address a range of societal challenges, enhance human well-being, and promote biodiversity with high effectiveness and economic efficiency (Sowińska-Świerkosz and García 2022).

In addition to their direct objectives, NbS generate a wide range of environmental, social, and economic co-benefits (Han and Kuhlicke 2021). Ecological co-benefits include regulating ecosystem services such as improved air, water, and soil quality, climate regulation, and carbon sequestration, while cultural services contribute recreational and tourism opportunities that sustain local economies. Socially, the co-design of NbS and the creation of green spaces can strengthen community cohesion and inclusion (Ommer et al. 2022). NbS also fosters resilience through circular economy strategies—reuse, reduce, recycle, remanufacture, and repurpose—that minimize waste and close material and energy loops (Stefanakis et al. 2021). These benefits influence stakeholder perceptions and are critical to successful implementation. However, limited awareness of such co-benefits and the challenge of monetizing ecosystem services often discourage investment. To address this, scholars have attempted to quantify their economic value; for instance, Yamada et al. (2022)estimated the annual value of the traditional Japanese *kasumetie* levee system at up to 43,830 JPY per 1000 m^2.

At the global level, two indicators can be used to gauge the potential of NbS in DRR: national necessity drivers and national capability indicators (Tyllianakis et al. 2022). The first measures exposure and susceptibility to climate-related risks, whereas the second evaluates a nation's institutional and systemic ability to implement NbS as adaptation and mitigation strategies. The comparative analysis indicates that high-scoring countries across both indicators are grouped together, irrespective of geographical location or economic status, and there is an indication that NbS can be equitable entry points for DRR. These results illustrate the potential of NbS to complement or replace grey infrastructure and to connect climate adaptation strategies in both the Global North and South contexts.

The conceptualization of ecosystem services—human benefits derived from nature—has long laid its focus on monetary valuation. More recently, the Nature's Contributions to People (NCP) concept expanded this understanding by including diverse knowledge systems and by considering both benefits and disbenefits (Díaz et al. 2018). This transformation emphasizes that NbS go beyond ecological services: they also foster a sense of belonging, empowerment, and stewardship in communities. It is here that the idea of eco-social resilience gets importance, as it encompasses the ability of ecosystems to enable communities to "bounce back better" by bolstering social cohesion and empowerment.

4.1.1 Frameworks and Key Concepts

Understanding eco-social resilience and the role of ecosystems in disaster risk reduction requires clarity on several interrelated concepts. Over the past two decades, terms such as NbS, Ecosystem-Based Disaster Risk Reduction (Eco-DRR), Ecosystem-Based Adaptation (EbA), Green Infrastructure (GI), Ecosystem

Services (ES), and Nature's Contributions to People (NCP) have emerged in policy and academic debates. While these frameworks share common ground, they differ in emphasis, scope, and application. This section outlines these foundational concepts, providing the lens through which global experiences and case studies in the later sections can be understood and compared.

The International Union for Conservation of Nature (IUCN) adopted in 2016 the definition of NbS as "*actions to protect, sustainably manage, and restore natural or modified ecosystems that effectively and adaptively address societal challenges and benefit human well-being and ecosystems at the same time*" (IUCN 2016).

The European Commission defines NbS as "*actions influenced and supported by nature that are cost-effective, provide environmental, social, and economic benefits at the same time, and help build resilience*" (European Commission 2015).

NbS are often described as "*win-win, low-regret, or no-regret*" options (Anderson et al. 2022), which can be classified into:

- Green NbS—vegetation-based interventions (e.g., urban forests, wetlands)
- Blue NbS—water-based interventions (e.g., mangrove restoration, sponge cities)
- Hybrid NbS—combinations of green/blue with engineered (grey) structures (Sowińska-Świerkosz and García 2022)

Globally, NbS are gaining recognition. Over 60% of the *Nationally Determined Contributions* under the Paris Agreement (2015) explicitly identified ecosystem-based adaptation (EbA) or NbS as climate strategies (Stefanakis et al. 2021). Europe is considered a global leader in NbS-enabled innovation and policy mainstreaming (Davies et al. 2021).

The key approaches that promote nature and ecology to counter climate change-induced disasters were mainly green infrastructure (GI) and ecosystem services (ES). Very recently, the concept of NbS, ecosystem-based disaster risk reduction (EbDRR) and ecosystem-based adaptation (EbA) emerged (Dhyani et al. 2020). Among these, the GI is a relatively older concept meant to control urban sprawl through enhancing the ecological network and integrating green space (Barbosa et al. 2019). It took the form of roof gardens, vertical gardens, open green spaces, public parks, etc., that support climate change mitigation, disaster resilience, and community well-being (Mabon 2019).

According to the Millennium Ecosystem Assessment (MEA), Eco System Services (ES) are the direct and indirect benefits people derive from ecosystems (Orimoloye et al. 2021).

These include:

- Provisioning services—food, water, timber, fuel
- Regulating services—climate regulation, flood control, water purification
- Cultural services—recreation, spirituality, heritage
- Supporting services—soil formation, nutrient cycling

Based on the domain where the ecosystem-based approaches act, there were several concepts. All such concepts like ecosystem-based adaptation (EbA), mitigation (EbM), disaster risk reduction (Eco-DRR) and Green Infrastructure (GI) now fall under the umbrella concept of NBS in the context of local DRR (Calliari et al. 2019). For the purpose of this particular paper, these terms are defined accordingly. In ecosystem-based adaptation (EbA) the climate risk reduction benefits of NbS were recognised, where the natural capital used by the people to adapt to climate change impacts (Munang et al. 2013). Eco-DRR strives for resilient development through the management, conservation, and restoration of ecosystems (Estrella and Saalismaa 2013). When the former one considers only climatic disasters, the latter includes both climatic and non-climatic events, even though both fundamentally aim at enhancing the resilience of people with a focus on ecosystem management (Dhyani et al. 2018). Also EbA considers plans for long-term changes in climate and considers future risks, including slow-onset events like sea level rise and temperature increase, where the Eco-DRR focuses on present-day risks and disasters and aims to reduce the immediate impacts of floods, landslides, and storms (McVittie et al. 2018).

4.2 Understanding Stakeholder Perceptions in Eco-Social Resilience

The effectiveness of NbS in disaster risk reduction is strongly shaped by how they are perceived by stakeholders since it can determine the levels of acceptance, willingness to participate, and long-term sustainability of interventions. Perception may be understood as the process of collecting, selecting, and interpreting signals about uncertain impacts of events, activities, or technologies, which in turn influences adaptive behaviour and decision-making.

Positive perceptions often emerge when stakeholders recognize the multiple co-benefits of NbS, such as improved landscape aesthetics, recreation opportunities, or enhanced social cohesion (Chou 2016; Han and Kuhlicke 2021). However, NbS also faces scepticism. Challenges such as the time lag in visible results, the novelty and complexity of NbS compared to conventional grey options, and perceived disservices—including pest outbreaks, allergens, or obstructed views—can lead to resistance, especially in urban settings (Anderson et al. 2022; Russo et al. 2017). It can also vary according to individual characteristics (age, gender, past experiences, expectations), social factors (awareness, participation, sense of responsibility), and contextual influences (previous hazard exposure, trust in institutions, cultural values) (Anderson and Renaud 2021).

Studies show that there were three dimensions for its successful implementation:

- *Stakeholder success*—human satisfaction and acceptance
- *Ecological success*—achievement of intended ecosystem functions
- *Learning success*—knowledge generation and transfer for future practice (Palmer et al. 2013)

Studies have identified that the people's positive attitudes towards Laojie river restoration are largely associated with landscape aesthetics and recreational value, rather than water quality and biodiversity (Chou 2016). Similarly Han and Kuhlicke (2019) also noted that the aesthetic and recreational aspects have influenced people more positively.

Consequently, there exists a significant necessity to comprehend community responses concerning the success of implementation, so that it is feasible to sustain a favourable perspective towards green infrastructure in risk reduction. Otherwise, the increase in NbS-related projects may be in vain, and reliance on grey infrastructure could resurface, as indicated by Anderson et al. (2022). This underscores the importance of integrating regular feedback monitoring. Consequently, the communication process for this transition must be grounded in evidence collection regarding the effectiveness of NbS in generating co-benefits. In this context, assessment frameworks gain significance, becoming a substantial part of the literature. Additionally, there is an effort to pre-assess potential co-benefits and disbenefits of NbS in addressing Disaster Risk Reduction, which is essential for the selection and approval process of NbS (Ommer et al. 2022). A participatory System Dynamics Model for evaluating benefits and co-benefits has been developed by Pagano et al. (2019). Given that stakeholder perceptions can influence the evaluation of co-benefits, Giordano et al. (2020) propose an innovative methodology for assessing the effectiveness of NbS by taking into account the variations in stakeholders' perceptions. The case study presented by Ruckelshaus et al. (2020) emphasizes the application of new data technologies and modelling frameworks to overcome a critical technical barrier in incorporating NbS into policy-relevant contexts: estimating the contribution of ecosystems to disaster risk reduction at regional and local decision-relevant scales.

The social dimension of NbS, being inherently rooted in the engagement of different stakeholders, is often an overlooked principle. Yet it has the potential to inculcate a sense of ownership and responsibility (Van Onselen et al. 2025) and also reduce conflicts of interest arising due to different degrees of perception. The community engagement is a gendered process mediated by individual values and their attachments to place and community (Lansakara et al. 2024). The prospect of adaptation actions leading to green, thriving worlds and mitigating global climate change has been seen as a driver, rather than solely economic factors. Yet, this driver remains open to gender discourse, with males generally driven by economic rewards. The motivational variations are divided into three clusters: Communitarian Adaptors, who are strongly motivated to adapt for the common good of society and nature, and who have a significant surplus of women; while they tend to be climate anxious and pro-environmental, they are less likely to take direct adaptation measures or work with local authorities. Individualist Adaptors, typically younger and of higher income (chiefly male), view themselves as strongly competent, preferring separate and technical adaptation efforts, are not anxious about climate change, and might be discouraged by intervention by authorities. On the other hand, Fatalist Adaptors are older in age, are climate change and official communications sceptics, have low adaptation motivation (especially ecological), and are rarely engaged in adaptation or pro-environmental activities (Brink and Wamsler 2019). Misconceptions regarding its effectiveness, lack of awareness, resistance, political orientation, and others may be a hindrance to participatory engagement among communities (Van Onselen et al. 2025).

4.3 Global Case Studies: Local Strategies for Eco-Social Resilience

The 2004 Indian Ocean tsunami was a watershed moment that revealed the critical role of ecosystems in reducing disaster risk. Since then, the global uptake of NbS has accelerated, although many of these approaches build on centuries-old practices such as crop calendars, blue–green infrastructure, and tank cascades (Ghaleh and Ghaleh 2020). While approaches vary widely across regions, the global evidence underscores recurring themes: urban climate adaptation, community-led engagement, ecosystem restoration for livelihoods, and indigenous and local knowledge systems.

4.3.1 Urban Climate Adaptation and Disaster Risk Reduction

The projection studies show that the urban population may reach more than half of the total world population by 2050, making it the battlefield of environmental and social issues. In response, urban regeneration has evolved, which aims to use urban spaces and social innovation to find solutions for social problems (dos Santos Figueiredo et al. 2022). If properly managed, it will provide many opportunities for a wide array of benefits to the community and to the city (Mukherjee et al. 2022). Literature advocates the integration of NBS into this urban regeneration, which is common nowadays to ensure its sustainability (Xiang et al. 2017).

- CLEVER Cities Project (Europe, China, and Ecuador): Implemented in Hamburg, London, and Milan (front-runner cities), with extensions to Quito (Ecuador) and Chinese cities, this project demonstrated how NbS can be integrated into urban regeneration and sustainable development strategies (NetworkNature 2023).
- GreenInfra4Beira (Mozambique): A program in Beira employed green infrastructure planning to reduce flooding, erosion, and heat stress while simultaneously creating employment, food security, and recreation opportunities (Letitre et al. 2015).
- Green Lungs of the City Project (China): In Yiwu City, Zhejiang Province, a forest–wetland ecological park enhanced biodiversity, water purification, and cultural services such as health and education benefits (NetworkNature 2023).

Compared to a normal flood event, urban flooding is a highly complex phenomenon, and its risk level is determined by both flooding characteristics and the nature of society, which need to be tackled differently (Alves et al. 2024). The *sponge city* initiatives in Climate-Resilient and Smart Urban Water Infrastructure Project of China are one among them. It is a stormwater management through increased infiltration, detention, storage, treatment, and drainage (Hamidi et al. 2021). The project is renowned for its gender action plan, which promotes women's participation in the workforce and decision-making processes, ensuring that resilience-building efforts benefit from diverse expertise. Rather than resilient infrastructure it builds resilient leadership through workforce inclusion, skill development and knowledge platform (Asian Development Bank 2025).

- "Sponge park" in Manchester: West Gorton Community Park in Manchester, England, functions as a 'sponge park,' designed to absorb and retain water using the same principles as a sponge city. It features a big play area with permeable safety surfacing and a basketball court with a non-permeable surface that redirects the rain to the drainage channel into a bio-retention tree-pit (Manchester City Council 2023).
- Room for the River (Netherlands): A large-scale initiative that redesignated floodplains, combining hydraulic safety with improved spatial quality, biodiversity, and cultural landscapes (Herk et al. 2012)
- Laojie River Restoration (Taiwan): This project transformed a flood-prone river in Taoyuan City into linear green infrastructure, improving landscape aesthetics and recreational values while reducing urban flood risk (Chou 2016).

4.3.2 Community Engagement and Social Inclusiveness

Although literature frequently asserts that empowering local communities in ecosystem-based disaster risk reduction (DRR) guarantees its effectiveness, sustainability, relevance, fairness, acceptance, and community ownership, such inclusiveness has often remained confined to documents only (Lansakara et al. 2024). Numerous examples of globally recognized practices exist that aim to incorporate community input and enhance the feeling of community ownership.

- Community-led NbS in Faridpur (Bangladesh) and Sialkot (Pakistan): It included green parks, retention ponds, and rooftop gardens. In Sialkot, a multifunctional park integrated ornamental trees, playgrounds, kiosks for small enterprises, solar-powered lighting, and rainwater collection systems, directly addressing flood risks while serving local livelihoods. The projects emphasized community ownership by drawing on community-level climate risk and vulnerability assessments, and traditional knowledge, as well as by co-designing measures that would be operated and maintained by the respective communities (United Nations Office for Disaster Risk Reduction and Group of Twenty 2024).

- Wetland management: Globally wetlands fall into three categories, viz., communal wetlands, and wetlands, belonging to private ownership and those that are protected areas. A study that aimed at comparing the state of various categories of wetlands in eastern Free State, South Africa has found that communal wetlands were more degraded, while wetlands in protected areas and in private commercial farms were in a good ecological state. This is attributed mainly to the lack of awareness regarding the mitigation effect of mangroves against various risks (Belle et al. 2018). Whereas in North Jakarta, Indonesia, the mangroves fall into two categories—protected and unprotected areas based on distinct management practices and conditions. The sustainability analysis has been identified that protected mangrove forests, which are under the supervision of the government, can enhance their full potential ecosystem services and DRR capacity, whereas the unprotected areas face challenges due to less effective management, which necessitated the involvement of community-based organisations and various other stakeholders in this regard (Setiacahyandari and Hizbaron 2024). So there is an urgent need for community-level mobilisation for its management so that the full DRR potential of each ecosystem service will emerge.

4.3.3 Ecosystem Restoration and Livelihoods

Ecosystem restoration not only rehabilitates degraded landscapes but also reduces disaster risk and supports livelihoods.

- Mangrove Restoration (Thailand, Philippines, Indonesia): Knowing the potential of mangroves, there have been massive restoration activities all over the world. Thailand is known for mangrove ecosystem restoration with the involvement of the community since 2005 (Datta et al. 2012). The National Greening Program (NGP) of the Philippines similarly aims at the restoration of mangrove ecosystems along the coasts to lessen the impact of natural hazards and enhance community resilience (Quevedo et al. 2020). Indonesia's *National Mangrove Restoration Programme (NMRP)* targets 600,000 hectares, sequestering 200–350 tons of carbon per hectare and potentially contributing 7–12% of Indonesia's national emissions reduction target (Setyanigrum et al. 2025).
- Community-Based Mangrove Management (South Asia): While CBMM initiatives have initially succeeded, sustainability often faltered due to donor withdrawal, internal inequality, and political capture, as seen in the Sundarbans, India (Datta et al. 2010, 2012). On the contrary, appropriate recognition of the requirements of multiple stakeholders in institutional decision-making through mixed land use and resource utilization methods could foster both ecological sustenance and economic prosperity, as exemplified in the case of the Philippines (Quevedo et al. 2020).

- Farmer-Managed Natural Regeneration (West Africa): FMNR encourages natural regrowth of trees on farmland, combining soil conservation, food production, and biodiversity protection in cost-effective ways (Weston et al. 2015).
- Kuttanad Farming System (Kerala, India): A below-sea-level rice–fish farming system recognized as a Globally Important Agricultural Heritage System (Food and Agricultural Organisation 2012). It balances aquaculture, agriculture, and flood management, exemplifying traditional NbS adapted to local hydrology (Food and Agriculture Organization of the United Nations 2012).

4.3.4 Indigenous and Local Knowledge Systems

Indigenous knowledge demonstrates that NbS are not new but deeply embedded in cultural practices that have historically managed risk.

- Japan: It is a country that has a long tradition of protecting natural resources to reduce risks that emerge from local knowledge. Kasumeti, yashkirin, waju, Nakatsu Tidal flat, etc., were some of the traditional approaches.
- *Kasumeti* is a discontinuous levee that has been described as a type of riverbank, which allows floodwater to flow into the surrounding paddy fields, which act as temporary flood control basins (Yamada et al. 2022). Creation of marshy lowland (*waju*) to protect against flood impact is another one. *Yashkirin* refers to homestead woodland or windbreak trees surrounding the houses to protect houses from wind and can be seen in many areas of Japan (the Ministry of the Environment, Japan 2016). For tidal protection, a combination of seawall and *Tidal flat* has been employed. Here the tidal flat acts as a natural wave attenuator where gentle slopes and shallow depths allow it to absorb wave energy, reducing the impact of waves on the coastline (Reed et al. 2018).
- The traditional agroforestry system of the Kunisaki Peninsula–Usa region in Japan—linking sawtooth oak forest management, shiitake mushroom cultivation, and irrigation ponds—serves as a sustainable model for integrated water-resource management. The Shiitake mushrooms are grown on logs cut from Sawtooth Oak trees. The spent oak trees will be returned to the forest itself so that it will enrich the soil by decomposing into organic matter, which can enhance the overall forest metabolism (Ministry of the Environment, Japan 2016).
- Leaky Dams (Europe and Global): Also called woody debris dams, these hybrid structures slow down water flow in catchments, reducing downstream flood risks while maintaining ecological functions (Hansamali et al. 2025).
- Sri Lanka: The ancient tank cascade system aligned with natural landscapes to manage floods and sustain agriculture, providing lessons for contemporary NbS (Hewawasam and Matsui 2022).
- China: The Hani rice terraces of Yunnan Province exemplify how ethnic communities utilized slopes and vegetation to manage water resources, creating a cultural landscape of resilience (Zhan and Jin 2015).

- Mali and the Great Green Wall: Decentralized forest governance in Segou brought 26,700 hectares under sustainable management, while the Great Green Wall program engages communities in regreening arid lands. This is supported by the NGO named Tree Aid that aims at eco-social resilience among people living in the drylands of Africa by helping them to overcome poverty and the effects of climate change by growing trees, improving incomes, and restoring land. Similarly the great green wall program in Mali, where working with communities to plant and regenerate native trees to regreen the area of Massantola in the Koulikoro region (Tree Aid n.d.).

4.4 Poverty Alleviation and Livelihood Protection

Establishment of NbS often provides both direct and indirect assistance to overcome various socio-economic hurdles through poverty alleviation and livelihood protection.

- In Japan's Kunisaki Peninsula Usa region, the cultivation of shiitake mushrooms on sawtooth oak logs has created diverse livelihood opportunities such as forestry, agriculture, and aquaculture. These activities not only reduce vulnerability to environmental and market shocks but also strengthen community resilience (Ministry of the Environment, Japan 2016).

It is high time to introduce some economic incentives, like indirect benefits, to NbS-related actions for reducing community discrimination and promoting their involvement. Also such market mechanisms can have an impact on social concerns like poverty alleviation (Dalimunthe 2018).

- Payments for Ecosystem Services (PES): PES offer financial incentives to individuals and communities for conserving ecosystem services and managing natural resources (Razzaque 2017). The case study from Tongke-Tongke, Indonesia, is a clear success story of bridging the nature of PES between economic development and environmental protection (Malik et al. 2025). Similarly in the case of the Tahiry Honko project from the Bay of Assassins, Southwest Madagascar, a PES scheme for reducing deforestation and degradation and restoring mangroves in the region has been established, which can reduce the emissions of 1300 tonnes of carbon dioxide per year. Here the PES is based on the carbon credits, which are generated by conserving this ecosystem, which can contribute towards poverty alleviation and social development, thus offering the community a fixed flow of income. Even though the lack of clear legal frameworks and coordination between the sectoral ministries, weakness of government organisations due to political instability and limited local governance capacity are the major challenges for the implementation of PES (payment for ecosystem services) schemes (Rakotomahazo et al. 2023).

- Mesoamerican Reef Insurance Programme: Mesoamerican Reef, the second-largest barrier reef in the world, provides many DRR services. As an effort to preserve these services, the Nature Conservancy, the Mexican state government of Quintana Roo and other partners established the Mesoamerican Reef Insurance Programme. It combines ecosystem-based adaptation measures with climate risk financing and insurance by providing insurance against tropical cyclones and covers damages to the coral reefs found in the Mesoamerican Reef, as well as the beaches along the coast. In this regard, it transfers the risk of damages from tropical storms and provides for the protection of vital ecosystems through coral reef protection, management and restoration, and the stabilization of beachfronts. The reef insurance is replenished through an existing fee levied from owners of beachfront properties and paid into a trust fund (Wagner et al. 2019).

4.4.1 Policy Integration and Institutional Governance

There were instances in which the state governance systems have incorporated the EcoDRR in policies, all of which have varying levels of community engagement provision. The case study from Srikakulam, Andhra Pradesh, India marked the incorporation of DRR and CCA but failed to develop an integrated strategy and offered limited avenues for engagement with local-level functionaries and communities or it was too centralised (Dash and Akhter 2023). In examining the mainstreaming of Eco-DRR within local-level development plans in rural South Africa, Mugari and Nethengwe (2022) observed that community engagement emerged as a consistent theme across all policies. But still the integration has been subjected to many challenges and a lack of genuineness. Many of the community meetings that aimed at incorporating their knowledge have been noted with low attendance mainly due to the inability of the ward committees to effectively engage with them, in addition to the unawareness of their actual roles and responsibilities as they act as a major link between municipality and community. Lack of knowledge, different levels of local awareness, as well as such internal factors as motivation, locus of control, responsibilities, and priorities, can impact their involvement. The lack of trust among the community has left them behind in the process, so it is important to ameliorate relations between them.

4.5 Conclusion

Traditionally, grey infrastructure has represented the default pathway for managing disaster risk. However, there is a growing shift towards NbS, driven mainly by their ability to produce multiple co-benefits in addition to risk reduction. NbS not only addresses existing hazards but also enhances quality of life through the provision of ecological, social, and economic services. Cultural services like co-design, recreation, and tourism directly promote social inclusion and social cohesion, while livelihoods support poverty reduction and better well-being.

Case studies from Japan, Sri Lanka, China, and Africa demonstrate how culturally rooted practices can be adapted for contemporary risk reduction. Integrating traditional wisdom with modern science enhances both ecological and social resilience. Restoration-oriented NbS function as "resilience engines," producing eco-social dividends such as climate mitigation, poverty reduction, and food security. Similarly, urban projects like sponge cities and river restoration demonstrate that disaster risk reduction and social development can advance in parallel, offering hybrid solutions that manage water while enhancing recreation, aesthetics, and biodiversity.

A central social dimension of NbS lies in community involvement in design and implementation. Active participation strengthens project effectiveness, transparency, and acceptance, while fostering a sense of ownership among beneficiaries. Yet in practice, community involvement often remains superficial and faces challenges such as limited knowledge, scepticism about NbS compared to grey options, internal conflicts, and power imbalances. Personal factors—age, gender, past experiences, and expectations—also shape perceptions, alongside broader concerns about novelty, complexity, and potential disservices of NbS, such as pests, allergies, or blocked views in urban areas. To sustain momentum, NbS initiatives must be supported by continuous communication, evidence-based monitoring, and transparent feedback mechanisms. Demonstrating their effectiveness and co-benefits is essential to building trust and avoiding a regression toward reliance on grey infrastructure. Without such efforts, the surge of NbS projects risks losing credibility, undermining their potential as equitable, sustainable strategies for disaster risk reduction and eco-social resilience.

References

Adarsh T, Haneena Sulthana KP (2025) Trends and insights in nature-based solutions: a bibliometric study. Int J Disaster Stud Clim Resil 1(1):22–34. https://resiliencepress.org/index.php/disaster/article/view/3

Alves PBR, Amanguah E, McNally D, Espinoza M, Ghaedi H, Reilly AC, Hendricks MD (2024) Navigating the definition of urban flooding: a conceptual and systematic review of the literature. Water Sci Technol 90(10):2796–2812. https://doi.org/10.2166/wst.2024.351

Anderson CC, Renaud FG (2021) A review of public acceptance of nature-based solutions: the 'why', 'when', and 'how' of success for disaster risk reduction measures. Ambio 50(8):1552–1573. https://www.jstor.org/stable/48739160

Anderson CC, Renaud FG, Hanscomb S, Gonzalez-Ollauri A (2022) Green, hybrid, or grey disaster risk reduction measures: what shapes public preferences for nature-based solutions? J Environ Manag 310:114727. https://doi.org/10.1016/j.jenvman.2022.114727

Asian Development Bank (2025) Women as agents of change: shaping resilience through sponge cities [case study]. Development Asia Retrieved 29 Aug 2025, from https://development.asia/case-study/women-agents-change-shaping-resilience-through-sponge-cities

Barbosa A, Martín B, Hermoso V, Arévalo-Torres J, Barbière J, Martínez-López J, Domisch S, Langhans SD, Balbi S, Villa F, Delacámara G, Teixeira H, Nogueira AJA, Lillebø AI, Gil-Jiménez Y, McDonald H, Iglesias-Campos A (2019) Cost-effective restoration and conservation planning in green and blue infrastructure designs. A case study on the Intercontinental Biosphere Reserve of the Mediterranean: Andalusia (Spain)—Morocco. Sci Total Environ 652:1463–1473. https://doi.org/10.1016/j.scitotenv.2018.10.416

Belle JA, Collins N, Jordaan A (2018) Managing wetlands for disaster risk reduction: a case study of the eastern Free State, South Africa. JÀMBÁ J Disaster Risk Stud 10(1). https://doi.org/10.4102/jamba.v10i1.400

Brink E, Wamsler C (2019) Citizen engagement in climate adaptation surveyed: the role of values, worldviews, gender and place. J Clean Prod 209:1342–1353. https://doi.org/10.1016/j.jclepro.2018.10.164

Calliari E, Staccione A, Mysiak J (2019) An assessment framework for climate-proof nature-based solutions. Sci Total Environ 656:691–700. https://doi.org/10.1016/j.scitotenv.2018.11.341

Chou R-J (2016) Achieving successful river restoration in dense urban areas: lessons from Taiwan. Sustainability 8(11):1159. https://doi.org/10.3390/su8111159

Dalimunthe SA (2018) Who manages space? Eco-DRR and the local community. Sustainability 10(6):1705. https://doi.org/10.3390/su10061705

Dash B, Akhter MS (2023) Mainstreaming disaster risk reduction (DRR)-climate change adaptation (CCA) and governance: Srikakulam, Andhra Pradesh, India. Int J Disaster Risk Reduct 86:103560. https://doi.org/10.1016/j.ijdrr.2023.103560

Datta D, Guha P, Chattopadhyay RN (2010) Application of criteria and indicators in community-based sustainable mangrove management in the Sunderbans, India. Ocean Coast Manag 53(8):468–477. https://doi.org/10.1016/j.ocecoaman.2010.06.007

Datta D, Chattopadhyay RN, Guha P (2012) Community-based mangrove management: a review on status and sustainability. J Environ Manag 107:84–95. https://doi.org/10.1016/j.jenvman.2012.04.013

Davies C, Chen WY, Sanesi G, Lafortezza R (2021) The European Union roadmap for implementing nature-based solutions: a review. Environ Sci Pol 121:49–67. https://doi.org/10.1016/j.envsci.2021.03.018

De Silva A, Amaratunga D, Haigh R (2022) Green and blue infrastructure as nature-based better preparedness solutions for disaster risk reduction: key policy aspects. Sustainability 14(23):16155. https://doi.org/10.3390/su142316155

Dhyani S, Lahoti S, Khare S, Pujari P, Verma P (2018) Ecosystem-based disaster risk reduction approaches (EbDRR) as a prerequisite for inclusive urban transformation of Nagpur City, India. Int J Disaster Risk Reduct 32:95–105. https://doi.org/10.1016/j.ijdrr.2018.01.018

Dhyani S, Karki M, Gupta AK (2020) Opportunities and advances to mainstream nature-based solutions in disaster risk management and climate strategy. In: Dhyani S, Gupta AK, Karki M (eds) Nature-based solutions for resilient ecosystems and societies. Springer, Singapore, pp 1–26. https://doi.org/10.1007/978-981-15-4712-6_1

Díaz S, Pascual U, Stenseke M, Martín-López B, Watson RT, Molnár Z, Hill R, Chan KMA, Baste IA, Brauman KA, Polasky S, Church A, Lonsdale M, Larigauderie A, Leadley PW, van Oudenhoven APE, Demissew S, Erpul G, Failler P et al (2018) Assessing nature's contributions to people. Science 359(6373):270–272. https://doi.org/10.1126/science.aap8826

dos Santos Figueiredo YD, Prim MA, Dandolini GA (2022) Urban regeneration in the light of social innovation: a systematic integrative literature review. Land Use Policy 113:105873. https://doi.org/10.1016/j.landusepol.2021.105873

Estrella M, Saalismaa N (2013) Ecosystem-based disaster risk reduction (eco-DRR): an overview. In: Renaud FG, Sudmeier-Rieux K, Estrella M (eds) The role of ecosystems in disaster risk reduction. United Nations University Press. https://www.google.co.in/books/edition/The_Role_of_Ecosystems_in_Disaster_Risk/9mTRjKmN1LwC?hl=en&gbpv=0

European Commission (2015) Towards an EU research and innovation policy agenda for nature-based solutions & re-naturing cities: final report of the horizon 2020 expert group on 'Nature-based solutions and re-naturing cities' (full version). Publications Office of the European Union. https://doi.org/10.2777/765301

Food and Agriculture Organization of the United Nations (2012) Kuttanad below sea-level farming system [case study]. FAO Family Farming Knowledge Platform. Retrieved 29 Aug 2025, from https://www.fao.org/family-farming/detail/en/c/283069/

Food and Agriculture Organization of the United Nations (2012) Kuttanad farming system—India GIAHS.https://www.fao.org/giahs/giahs-around-the-world/india-kuttanad-farming-system/en

Ghaleh MR, Ghaleh MR (2020) The role of smart flood management in the ancient blue-green infrastructure. In: 2020 IEEE International Symposium on Technology and Society (ISTAS). IEEE, pp 417–421. https://doi.org/10.1109/ISTAS50296.2020.9462209

Giordano R, Pluchinotta I, Pagano A, Scrieciu A, Nanu F (2020) Enhancing nature-based solutions acceptance through stakeholders' engagement in co-benefits identification and trade-offs analysis. Sci Total Environ 713:136552. https://doi.org/10.1016/j.scitotenv.2020.136552

Hamidi A, Ramavandi B, Sorial GA (2021) Sponge City—an emerging concept in sustainable water resource management: a scientometric analysis. Resour Environ Sustain 5:100028. https://doi.org/10.1016/j.resenv.2021.100028

Han S, Kuhlicke C (2019) Reducing Hydro-Meteorological Risk by Nature-Based Solutions: What Do WeKnow about People's Perceptions? Water, 11(12):2599. https://doi.org/10.3390/w11122599

Han S, Kuhlicke C (2021) Barriers and drivers for mainstreaming nature-based solutions for flood risks: the case of South Korea. Int J Disaster Risk Sci 12(5):661–672. https://doi.org/10.1007/s13753-021-00372-4

Hansamali U, Makumbura RK, Rathnayake U, Azamathulla HM, Muttil N (2025) Leaky dams as nature-based solutions in flood management part I: introduction and comparative efficacy with conventional flood control infrastructure. Hydrology 12(4):Article 4. https://doi.org/10.3390/hydrology12040095

Herk S, van, Rijke J, Zevenbergen C, Ashley R (2012) Transition in governance of river basin managementin the Netherlands through social learning [Conference presentation]. Int Sustain TransitS, Copenhagen, Denmark

Hewawasam V, Matsui K (2022) Traditional water governance practices for flood mitigation in ancient Sri Lanka. Climate 10(5):Article 5. https://doi.org/10.3390/cli10050069

International Union for Conservation of Nature (2016) Resolution 69: defining nature-based solutions (WCC-2016-Res-069). In: IUCN resolutions, recommendations and other decisions, World Conservation Congress, 6–10 September 2016. IUCN. https://portals.iucn.org/library/sites/library/files/resrecfiles/WCC_2016_RES_069_EN.pdf

Lansakara DC, Le De L, Petterson M, Wickramasinghe D (2024) The potential for community-driven ecosystem-based disaster risk reduction in South Asia: a literature review. Disaster Prevent Manage Int J 33(2):78–97. https://doi.org/10.1108/DPM-06-2023-0128

Letitre P, Brils J, Van de Ven FHM, Schipper CA, Brolsma R, Snep R, Aalbers C, Simon C, Krijgsman A, Klein J, Mondée H, Blommaert A, Lansbergen F, Verwey A, Eucker D, Niquisse S, Chissone Lovane G, Céjar A (2015) GreenInfra4Beira. Technical report, Deltares. https://doi.org/10.13140/RG.2.1.3169.3285

Mabon L (2019) Enhancing post-disaster resilience by 'building back greener': evaluating the contribution of nature-based solutions to recovery planning in Futaba County, Fukushima Prefecture, Japan. Landsc Urban Plan 187:105–118. https://doi.org/10.1016/j.landurbplan.2019.03.013

Malik A, Ichsan Ali M, Rasyid Jalil A, Mannan A, Musyawarah R (2025) Promoting sustainable mangrove tourism through payments for ecosystem services: insights from Tongke-Tongke Village, South Sulawesi, Indonesia. Regional Sustain 6(2):100213. https://doi.org/10.1016/j.regsus.2025.100213

Manchester City Council (2023) Manchester's West Gorton climate-resilient 'sponge park' [case study]. UK100. Retrieved 29 Aug 2025, from https://uk100.gn.apc.org/projects/knowledgehub/manchesters-sponge-park

McVittie A, Cole L, Wreford A, Sgobbi A, Yordi B (2018) Ecosystem-based solutions for disaster risk reduction: lessons from European applications of ecosystem-based adaptation measures. Int J Disaster Risk Reduct 32:42–54. https://doi.org/10.1016/j.ijdrr.2017.12.014
Ministry of the Environment, Japan (2016) Ecosystem-based disaster risk reduction in Japan: a handbook for practitioners. Nature Conservation Bureau, Ministry of the Environment. https://www.env.go.jp/en/nature/npr/eco-drr/handbook.html
Mugari E, Nethengwe NS (2022) Mainstreaming ecosystem-based disaster risk reduction: towards a sustainable and just transition in local development planning in rural South Africa. Sustainability 14(19):12368. https://doi.org/10.3390/su141912368
Mukherjee M, Wickramasinghe D, Chowdhooree I, Chimi C, Poudel S, Mishra B, Ali ZF, Shaw R (2022) Nature-based resilience: experiences of five cities from South Asia. Int J Environ Res Public Health 19(19):11846. https://doi.org/10.3390/ijerph191911846
Munang R, Thiaw I, Alverson K, Mumba M, Liu J, Rivington M (2013) Climate change and ecosystem-based adaptation: a new pragmatic approach to buffering climate change impacts. Curr Opin Environ Sustain 5(1):67–71. https://doi.org/10.1016/j.cosust.2012.12.001
Murti R, Mathez-Stiefel S (2019) Social learning approaches for ecosystem-based disaster risk reduction. Int J Disaster Risk Reduct 33:433–440. https://doi.org/10.1016/j.ijdrr.2018.09.018
NetworkNature (2023) Clever cities project wraps up 5-year initiative, marking milestones in urban regeneration. NetworkNature. Retrieved from https://networknature.eu/networknature/article/clever-cities-project-wraps-5-year-initiative-marking-milestones-urban-regeneration
Ommer J, Bucchignani E, Leo LS, Kalas M, Vranić S, Debele S, Kumar P, Cloke HL, Di Sabatino S (2022) Quantifying co-benefits and disbenefits of nature-based solutions targeting disaster risk reduction. Int J Disaster Risk Reduct 75:102966. https://doi.org/10.1016/j.ijdrr.2022.102966
Orimoloye IR, Zhou L, Kalumba M, A. (2021) Drought disaster risk adaptation through ecosystem services-based solutions: way forward for South Africa. Sustainability 13(8):4132. https://doi.org/10.3390/su13084132
Pagano A, Pluchinotta I, Pengal P, Cokan B, Giordano R (2019) Engaging stakeholders in the-assessment of NBS effectiveness in flood risk reduction: A participatory System Dynamics Model for benefits andco-benefits evaluation. Sci Total Environ, 690:543–555. https://doi.org/10.1016/j.scitotenv.2019.07.059
Palmer RN, Cardwell HE, Lorie MA, Werick W (2013) Disciplined planning, structured participation, and collaborative Modeling—applying shared vision planning to water resources. JAWRA Journal of the American Water Resources Association 49(3):614–628. https://doi.org/10.1111/jawr.12067
Quevedo JMD, Uchiyama Y, Kohsaka R (2020) Perceptions of local communities on mangrove forests, their services and management: implications for eco-DRR and blue carbon management for Eastern Samar, Philippines. J Forest Res 25(1):1–11. https://doi.org/10.1080/13416979.2019.1696441
Rakotomahazo C, Ranivoarivelo NL, Razanoelisoa J, Todinanahary GGB, Ranaivoson E, Remanevy ME, Ravaoarinorotsihoarana LA, Lavitra T (2023) Exploring the policy and institutional context of a payment for ecosystem services (PES) scheme for mangroves in southwestern Madagascar. Mar Policy 148:105450. https://doi.org/10.1016/j.marpol.2022.105450
Razzaque J (2017) Payments for ecosystem services in Sustainable Mangrove Forest Management in Bangladesh. Transnatl Environ Law 6(2):309–333. https://doi.org/10.1017/S2047102516000376
Reed D, van Wesenbeeck B, Herman PMJ, Meselhe E (2018) Tidal flat-wetland systems as flood defenses: understanding biogeomorphic controls. Estuar Coast Shelf Sci 213:269–282. https://doi.org/10.1016/j.ecss.2018.08.017
Renaud FG, Sudmeier-Rieux K, Estrella M (eds) (2013) The role of ecosystems in disaster risk reduction. United Nations University Press, Shibuya
Ruckelshaus M, Reguero BG, Arkema K, Compeán RG, Weekes K, Bailey A, Silver J (2020) Harnessing new data technologies for nature-based solutions in assessing and managing risk in coastal zones. Int J Disaster Risk Reduct 51:101795. https://doi.org/10.1016/j.ijdrr.2020.101795

Russo A, Escobedo FJ, Cirella GT, Zerbe S (2017) Edible green infrastructure: an approach and review of provisioning ecosystem services and disservices in urban environments. Agric Ecosyst Environ 242:53–66. https://doi.org/10.1016/j.agee.2017.03.026

Setiacahyandari HK, Hizbaron DR (2024) Understanding eco-DRR as a sustainability indicator for mangrove conservation in urbanized area of North Jakarta, Indonesia. Environ Sustain Indic 24:100494. https://doi.org/10.1016/j.indic.2024.100494

Setyanigrum W, Koesrianti, Narwati E (2025) Effectiveness of the National Mangrove Restoration Program (NMRP) in achieving net zero emission (NZE) targets in Indonesia. E3S Web Conf 622:02008. https://doi.org/10.1051/e3sconf/202562202008

Sowińska-Świerkosz B, García J (2022) What are Nature-based solutions (NBS)? Setting core ideas for concept clarification. Nat Based Solut 2:100009. https://doi.org/10.1016/j.nbsj.2022.100009

Stefanakis AI, Calheiros CSC, Nikolaou I (2021) Nature-based solutions as a tool in the new circular economic model for climate change adaptation. Circ Econ Sustain 1(1):303–318. https://doi.org/10.1007/s43615-021-00022-3

Tree Aid (n.d.) Local governance of forest resources in Mali (DAREX009). Biodiversity Challenge Funds. https://www.biodiversitychallengefunds.org.uk/project/DAREX009

Tyllianakis E, Martin-Ortega J, Banwart SA (2022) An approach to assess the world's potential for disaster risk reduction through nature-based solutions. Environ Sci Pol 136:599–608. https://doi.org/10.1016/j.envsci.2022.07.021

United Nations Office for Disaster Risk Reduction & Group of Twenty (G20) (2024) Good practices for increasing the application of nature-based solutions and ecosystem-based approaches for disaster risk reduction [PDF]. https://www.undrr.org/publication/good-practices-increasing-application-nature-based-solutions-and-ecosystem-based

Van Onselen VM, Ota M, Pratiwi ES, Lin WJ, Lin T-Y (2025) Community involvement in ecosystem-based disaster risk reduction: a scoping review to guide assessment metrics. IOP Conf Series Earth Environ Sci 1443(1):012010. https://doi.org/10.1088/1755-1315/1443/1/012010

Wagner N, Alderman HH, Alderman C (2019) Concept note for MAR Fund's Reef Rescue Initiative: Opportunities for the Mesoamerican Reef countries to include coral reefs into asset valuation and climate-related risk assessments, and the economic rationale for using financial instruments to insure coral reefs against hurricanes and damage from ship groundings. Prepared on behalf of MAR Fund, with the cooperation of CCAD and KfW. MAR Fund. https://marfund.org/en/wp-content/uploads/2020/02/Climate-Risk-Assessment-Concept-Note-Document.pdf

Weston P, Hong R, Kaboré C, Kull CA (2015) Farmer-managed natural regeneration enhances rural livelihoods in dryland West Africa. Environ Manag 55(6):1402–1417. https://doi.org/10.1007/s00267-015-0469-1

Xiang P, Wang Y, Deng Q (2017) Inclusive nature-based solutions for urban regeneration in a natural disaster vulnerability context: a case study of Chongqing, China. Sustainability 9(7):Article 7. https://doi.org/10.3390/su9071205

Yamada Y, Taki K, Yoshida T, Ichinose T (2022) An economic value for ecosystem-based disaster risk reduction using paddy fields in the kasumitei open levee system. Paddy Water Environ 20(2):215–226. https://doi.org/10.1007/s10333-021-00885-0

Zhan G, Jin Z (2015) Hani rice terraces of Honghe—the harmonious landscape of nature and humans. Landsc Res 40(6):655–667. https://doi.org/10.1080/01426397.2015.1060299

Chapter 5
Contours of Sociological Research on Climate Change and Beyond

K. P. Arunjith and Lekshmi V. Nair

Abstract Climate change is one of the most important environmental issues of the twenty-first century. Developing countries and economies in transition contribute to the hike in greenhouse gas emissions, which eventually culminates in climate change. On the other hand, indigenous communities worldwide conserve a significant share of the biodiversity, which helps regulate the causes of climate change. The paradox is that climate change effects significantly impact indigenous communities' lives, such as that of the tribal, fishing, and farming communities. Social science studies on human dimensions of climate change are still lagging, especially from a sociological perspective. This chapter explores the scope of sociology in studying climate change.

Keywords Climate change · Indigenous communities · Social science · Sociology · Anthropology

5.1 Introduction

We live in a changing climate scenario and have experienced recent extreme climatic events in and around us. World Economic Forum (WEF) identified ten crises in the world today; five of which are related to climate change, those are water pollution, air pollution, inundation, health issues and vector-borne diseases. Climate change debates have much importance in academia. Social science looks at the climate change debate with much academic interest.

Due to the Greenhouse Effect, the temperature near the Earth's surface increases. This phenomenon is called global warming, which leads to climate change.

K. P. Arunjith (✉) · L. V. Nair
Department of Humanities and Social Sciences, Indian Institute of Space Science and Technology, Thiruvananthapuram, Kerala, India
e-mail: arunjithkp.23@res.iist.ac.in; lvnair@iist.ac.in

J. K. Joseph, A. P. Pradeepkumar (eds.), *Eco-Social Work for Disaster Risk Reduction*, Disaster Resilience and Green Growth, https://doi.org/10.1007/978-981-95-4885-9_5

According to official records, Earth's average surface temperature is 15.5 °C with a rate of increase of 0.1 °C/decade. IPCC says the global mean temperature has risen by 1.09 °C since pre-industrial times (1850). Also, it is expected to increase between 1.5 and 5.8 °C by 2100. Global warming will exceed 3 °C by 2100, even if all commitments in the 2015 Paris Agreement are met. The UN chief (2023) said that July 2023 was the hottest month ever, signaling that the "era of global boiling has arrived." Temperature rise will not be uniform across the planet. Large changes in temperature will greatly accelerate other changes, such as changing patterns of rainfall and increasing temperature extremes.

According to UNDP (2023), climate is the average weather pattern in a specific area over a longer period, usually 30 or more years, that represents the overall state of the climate system. The UN defines climate change refers to long-term shifts in temperatures and weather patterns. According to NASA, climate change is a long-term change in the average weather patterns that have come to define Earth's local, regional and global climates.

5.2 Climate Change and Anthropocene

During the mid-twentieth century, anthropogenic activity, primarily the combustion of fossil fuels, elevated the concentration of heat-retaining greenhouse gases in the atmosphere, increasing the planet's average surface temperature. In 2002, Paul Crutzen asserted that humanity had transitioned from the Holocene to a new epoch known as the Anthropocene (Zalasiewicz et al. 2008). The new epoch represents a human-induced stratum in the geological record, distinguished by the worldwide effects of human activity on Earth Systems. The era which commenced in 1712, marked by the introduction of the steam engine, extended to 1850, when atmospheric indicators of industrialization became evident. The new man-made stratum significantly affected the environment. Humans have emerged as the most dominant species on Earth, instigating considerable global warming and alterations to land, environment, water, creatures, and the atmosphere.

5.3 Climate Change and People

The Declaration of Indigenous Peoples on Climate Change (2002) asserts that the origins of climate change stem from the production and consumption patterns of industrialized nations, which bear the main responsibility for the issue. The policies of developing nations and transitioning economies that endorse coal and uranium mining, logging, nuclear and large hydroelectric power generation, as well as oil and gas extraction and transportation, exacerbate climate change and the degradation of our lands (Declaration of Indigenous Peoples on Climate Change 2002, p. 99). All regions worldwide are currently encountering extreme weather and climate conditions. With the Earth's warming, scientists predict a rise in the frequency and severity of heat waves, flooding, precipitation, droughts, and cyclonic impacts on agricultural production in numerous locations globally.

According to the Sustainable Development Goals Report (2022), increased heat waves, droughts, and floods caused by climate change already affect billions of people worldwide. They are causing potentially irreversible changes in global ecosystems. In addition to that, famine, extreme poverty, sickness, the loss of fundamental resources in our historic habitats, are forcing the relocation of our people as environmental refugees. "The scientific evidence is unequivocal: climate change is a threat to human well-being and the health of the planet," the most recent report from the Intergovernmental Panel on Climate Change (IPCC) (2023) states. A short and quickly closing opportunity to ensure a livable future will be lost if coordinated global action is delayed any longer. There is food insecurity and hunger in portions of Africa, Central America, and South America as a result of floods and droughts brought on by climate change.

Everyone is impacted by climate change, but those who are most at risk are most severely affected. Between 3.3 and 3.6 billion people, according to the IPCC study (2023), reside in areas that are extremely vulnerable to climate change. The Arctic, Southern Asia, Central and South America, small island developing states, and a large portion of sub-Saharan Africa are hotspots of high human vulnerability. Humanitarian disasters that could force millions of people from their homes are caused by poverty, poor governance, conflict, and inadequate access to essential services. An estimated 700 million people could face displacement due to drought alone by 2030 (UN/World Bank High-Level Panel on Water 2018).

The United Nations Environment Programme (UNEP) (2023) has warned that those who depend heavily on farming, fishing, or forestry may lose their livelihoods as a result of climate change, and the IPCC has admitted that people with low incomes will be disproportionately affected. Numerous indigenous people inhabit delicate ecosystems, and their lands, resources, hunting, and fishing activities are likely to be impacted by climate change. This will impact their economies, health, and land (Declaration of Indigenous Peoples on Climate Change 2002, p. 98). The SDGs' emphasis on issues like education, poverty, access to justice, and climate change is particularly pertinent to indigenous communities (Magni 2017, p. 437). Many of the indigenous peoples inhabit the most endangered ecosystems. Adverse ecological phenomena, especially climate change, affect their land and result in biodiversity loss. These phenomena affected their social and cultural facets. For example, traditional indigenous practices and rituals have been disrupted by the significant loss of fauna and flora, and the growing scarcity of freshwater and food has resulted in health problems, particularly in children (Magni 2017). According to a paper published in *Nature*, India's biodiversity hotspots, which account for a sizeable fraction (16.86%) of the entire global area designated as such, are severely impacted by climate change (Chitale and Behera 2014).

The greatest significant environmental concern of the twenty-first century is global climate change. Hence studying the human dimension of climate change from social science perspectives opens new avenues. Nagel et al. (2010) rightly point out that the fundamental questions faced by a sociologist when embarking on climate change research are whether their research will be helpful to communities and policymakers seeking to mitigate or adapt to climate change, and whether such utility should be a high priority in identifying research questions and designing research programs.

5.4 Politics of Climate Change

Sustainability was defined as meeting the needs of the present without compromising the ability of future generations to meet their own needs (United Nations Brundtland Commission 1987). Due to the changing climate, sustainability is questionable. Also, social scientists are not prepared to spread a hopeful view on climate change. Because, whatever the mitigation measures implemented to lower greenhouse gas emissions, climate change will occur in the future (Bhatasara 2015). He attempts to make the case that governments, environmental activists, civil society, and global policy-level bodies are concerned about the extraordinary rise in global temperatures and that mitigation and adaptation alone will not be enough to stop the current climate change.

Climate change is creating new patterns of social and economic conflict, threatening the state's economic resources, increasing social inequality, and altering community structures (Norgaard 2018). Large-scale emission reductions are necessary to combat climate change, yet our petroleum-based economy is ingrained in our current political-economic structure. We need to act and participate democratically (Norgaard 2018). Norgaard analyzes climate skeptic misinformation campaigns, draws sociologists' attention to the ways that relationships between federal governments and oil company executives shape climate policy and public perception, and explains how even the idea of "balanced" media framing itself distorts public perception (Boykoff 2008; Norgaard 2018).

Communities also frequently engage in socially organized denial of climate change. Norgaard uses a Norwegian rural village as an example. Norgaard used the term "socially organized denial" to describe how disregarding knowledge about global warming occurs in reaction to social situations and is accomplished through social interaction. Norway maintains its economic interests by collectively ignoring climate change because oil production is so intimately tied to the nation's economic performance (Bhatasara 2015). Although climate change is a worldwide issue, there are several levels at which to look for solutions: local, regional, national, and international (Nagel et al. 2010).

5.5 Climate Change: A Matter of Inequality and Justice

The individuals who are most impacted by climate change are not the ones who produced it, making it a complex subject that involves global inequity and social justice. A sociological intervention is essential in this context. Sociological understanding of how global inequality manifests itself is therefore crucial, especially in emerging nations' local contexts (Bhatasara 2015).

According to Harlan et al (2015), there are three reasons why climate change is a matter of justice: (1) Unsustainable levels of greenhouse gas emissions are caused by social inequality; (2) the rich and the poor are affected by climate change in different ways; and (3) policies intended to manage climate change not only have different effects but also disproportionately exclude "the poor and the powerless." They point out that poorer countries are less accountable for the environmental problems of today and that wealthier countries consume "more than three times their share" of CO_2 emissions.

The authors define climate justice using the following four tenets: (1) "equity in distributing the burdens and sharing the benefits" of climate change; (2) participation by marginalized groups in the "governance and management of climate change"; (3) the freedom and capacity to make decisions that impact one's life; and (4) the restoration of historically damaged relationships, "correcting past wrongs against humanity, and restoring the Earth" are all components of climate justice. "Who has the right to control the Earth's common space?" the writers question. They say that any effort to lessen the effects of climate change must address the fundamental problems of global capitalism and that marginalized people, not only rich parties, should be able to participate in debates. They conclude that although sociology as a field has been crucial in tackling injustices and inequalities, interdisciplinary collaboration is required to help advance a climate change agenda (Brulle 2015, p. 186).

On a deeper level, sociology encourages us to think about how corporate oil interests, which profit from ongoing high emission scenarios, may subtly threaten democratic structures when they influence public opinion and policy discussions (Brulle et al. 2012; Dunlap 2013; Norgaard 2018). Also, the legitimacy of our current political and economic structures is at risk due to climate change (Klein 2014; Norgaard 2018). At the grassroots level, Walker (2011) noted that the role of the community in adaptation is crucial in the study of climate change. Delanty and Strydom (2003), sociologists, refer to the importance of community adaptation to climate change.

5.6 Social Science and Climate Change

The IPCC and NAS assert that global climate change is both a social and a physical concern. Technical solutions alone will not resolve the complex social, cultural, and behavioral transformations required for individuals to effectively confront the impending environmental issues (Nagel et al. 2010). Here the scope of social science is explicit.

In general, sociological as well as anthropological research on climate change lags behind other disciplines. Many scientists addressed the issue in different ways. Bhatasara (2015) clearly states that sociology is lagging behind, especially in studying the social and cultural aspects of climate change. Sociologists are inadequately represented in global climate change dialogues.

The argument raised by Norgaard is that subjects other than sociology, like psychology and economics, have conducted more studies on climate change than sociology because of their compatibility with the existing political and economic paradigms. Compared to sociologists, researchers in the fields of economics, psychology, and science communication have made the people's reaction to climate change a higher priority on their study agendas; their primary focus is mitigation. Among them, psychology and economics are more suited to the scientific models due to their compatibility with the political and economic paradigms that are already in place, as well as the fact that they employ humans as the unit of analysis (Brulle 2015; Norgaard 2018).

As we already mentioned, the significance of indigenous communities, especially those who are dependent on fishing, farming, and forestry in the wake of climate change, should be the subject matter of social science. Sociology, as well as anthropology, has a lot to contribute in this direction.

Nevertheless, sociological research on global climate change has been trailing behind studies in other social science disciplines on the same topic. However, the previous several decades have observed alterations in such trends. Social scientists assert that the role of social science is to enhance society's reflexive ability to tackle climate change, rather than merely providing advice to political entities (Brulle 2015).

According to Norgaard, sociology highlights the necessity of more significant structural adjustments to political, cultural, and economic systems. "The human-environment interface should be recognized by sociology as the primary social issue of the Anthropocene Age" (Brulle 2015). Furthermore, Grundmann et al. (2012) proposed that social behavior changes environmental processes, which in turn change social connections, regardless of whether we wish to protect society from the climate (adaptation) or mitigate it (mitigation) (Bhatasara 2015).

5.7 Sociology and Climate Change Research

Sociologists generally define climate change as a political and social phenomenon (Bhatasara 2015). Even though the central question raised from a sociological perspective is, since sociology tries to understand social facts, how will climate change be studied? Here, the two broader research fields in the human dimension of climate change emerge: adaptation and mitigation. Market-driven social structure leads to global warming and climate change. However, mitigation efforts should concentrate on those social structures. Again, climate change impacts directly or indirectly on communities, so communities, by themselves or with the help of civil society, develop different realms of adaptation.

Sociologists have always been skeptical of naturalistic explanations and have also been wary of teleological claims that provide conclusive future trajectories, because sociology has historically maintained a mostly reserved attitude about climate change (Lever-Tracy 2010). These claims always overlook the reality behind natural phenomena.

The nature-society gap is significant in sociology; Scott contends that the discipline must surmount this obstacle. Thus, highlighting the interplay between society and nature indicates that sociology should not consider alterations in nature as the primary basis for examining climate change; instead, it should focus on how climate change is co-constructed by the dynamic interactions of social and natural processes (Lidskog et al. 2014; Bhatasara 2015). Bhatasara asserts that climate change illustrates the dialectical relationship between nature and society, making it essential to sociology.

Sociologists faced a dilemma as climate change disrupts the fundamental principles of sociology. The discipline posits that society is a social construct rather than a natural determination. Characterizing climate change as a social construct does not diminish its truth or importance; scientists view this viewpoint as a form of denial and hindrance. This has placed sociologists in a vulnerable position. Brechin (2008) contends that it is necessary to transcend the conventional notion that just social facts precipitate social events (Bhatasara 2015).

Sociological research on climate change is currently underdeveloped due to challenges such as the disruption of foundational principles, issues related to constructivism, reduced financing for social research, and the politicization of the climate change discourse, which hinder sociology's progress in this area.

Sociology's myopia is rooted in historical factors. The discipline arose during the peak of the modernist belief that humanity had transcended natural "limits". The natural world was assumed to no longer significantly impact social outcomes (Brulle 2015; Norgaard 2018). The scientific community grapples with understanding social structure; it is imperative for sociologists to cultivate an ecological imagination (Norgaard 2018).

5.8 Sociology for Climate Change Research

Studies on climate change in sociology fall under the domain of environmental sociology (Nagel et al. 2010). Initially, environmental sociology focused on the social and political processes of the environmental movement. By the 1980s, it expanded its focus to examine the organizational, economic, cultural, and emotional elements affecting modern industrial society's interaction with the biophysical environment. This involved examining the environmental consequences of various economic development models, political disputes, prevailing disparities, and sustainability challenges (Nagel et al. 2010). Initially, environmental sociologists concentrated on the human-nature relationship; however, their study focus has recently shifted to global climate change (Nagel et al. 2010). Sociologists employ several key theoretical and methodological strategies to comprehend the interplay between humans and their natural environment, as well as to discern the ecologically significant characteristics of contemporary industrial nations and their influence on global climate change.

The importance of the two different types of imagination that a researcher in the field of sociology needs to employ in order to address climate change and the Anthropocene is explained by Norgaard (2018). (1) Ecological imagination, which is defined as the ability to detect the links between human actions and their impacts on the Earth's biophysical systems, and (2) sociological imagination, which is defined as the capacity to recognize the societal interactions that comprise this environmentally damaging social structure, are both required in order to address the problem of environmental degradation. He contended that "in the absence of a sociological imagination, the majority of individuals within both the scientific community and the general populace presently lack the capacity to perceive the social structure that envelops us." Furthermore, the central focus of sociology is to elucidate the interconnections among the "micro," "meso," and "macro" levels of social order, particularly in the context of climate change (Norgaard 2018).

Sociological research focuses on anthropogenic drivers of climate change, climate-related markets and organizations, consumption trends, global inequality and climate justice, strategies for adaptation, mitigation, response, and resilience, international climate politics, civil society and social movements, and public perception. Islam and Kieu (2021) delineate four comprehensive themes that provide a sociological perspective on climate change: effects of climate change; equitable mitigation and adaptation strategies; socio-political dynamics of advocacy; and public opinion, social theory, and methodology. Nagel et al. highlighted the human dimensions of climate change, such as the social causes, the social impacts, the social dimensions of mitigation and adaptation, and suggestions for enhancing sociological research. Norgaard (2018) asserted that the research on impact addresses inequalities related to race, gender, socioeconomic status, age, and disparities between developed and developing nations. Sociological research has examined class, gender, ethnicity, sexuality, and geography (Norgaard 2018).

Nagel et al. (2010) reiterate the NRC's proposed five domains for climate change mitigation and adaptation: environmental decision-making; environmental governance through the establishment of institutions; business decision-making with environmental considerations; individual behavior; and developing decision-relevant science consumption policies. He underlines that this research is aligned with sociology's theoretical and empirical strengths. In addition, climate research aims to understand how informal and formal entities, economic and governmental institutions, and NGOs shape, implement, and enforce climate policy.

According to Sunderlin (1995), there is not much sociological literature on how to adapt to climate change. What little there is, it primarily focuses on three areas: structural economic changes, managerial methods, and changes in culture and behavior, which broadly rely on Marxist, Weberian and Durkheimian traditions, respectively. The study says that the field of sociology of climate change adaptation needs to be widened and made more solid and dynamic so that it can better reflect how structures, processes, and conditions change in different places and how they affect adaptation. Another goal of sociology is to study how adaptation practices are organized in society by looking at how they start, spread, negotiate, and fight within social groups and how they become established (Bhatasara 2015).

Social thinkers contend that mitigation and adaptation help to reduce climate change's impacts. Mitigation focuses on the reduction of greenhouse gases, while adaptation pertains to human ability to coexist with the changing climate with alteration. While addressing the strategies for mitigation and adaptation, adaptability has a threefold structural, institutional, and societal. Researchers advocate for a hybrid policy approach to assess mitigation, which would encompass acknowledging global disparities, comprehending past impacts, and interrogating prevailing assumptions.

5.9 Theoretical Underpinnings

Bhatasara identifies the sociological subfields addressing climate change as Environmental Sociology, Political Sociology of Climate Change, Post-Carbon Sociology, and Risk Society, along with Structure-Agency Theories.

In the context of climate change, environmental sociologists have addressed the issue of "environmentalism" as a social movement. They have also been widely attacked for embracing a realistic perspective on climate change (Dunlap and Brulle 2015; Bhatasara 2015). According to Giddens (2009), there is still no unified political framework capable of effectively confronting climate change, and conventional politics fall short in addressing such complex challenges. To tackle this gap, he proposes a series of new ideas and carefully explores the link between climate change and energy security. Yet, Grundmann and Stehr (2010) argue that Giddens makes minimal use of sociological theory, including his own concepts of structuration and risk society. Consequently, his analysis highlights the nation-state as the key actor in carrying out both mitigation and adaptation measures. Lever-Tracy (2010) further points out that Giddens prioritizes political solutions over social and cultural forms of adaptation. In line with this, Bhatasara (2015) notes that Giddens' work is primarily intended for politicians, policymakers, administrators, and other professionals directly involved in climate change politics.

Urry (2009) explores the prospects of a future society that is expected to operate with lower carbon emissions. He argues that both institutions and processes have contributed to climate change, and therefore, adaptation efforts must address these dimensions together. Mulligan (2014) highlights that Urry takes an optimistic stance on the use of scenario-building and back-casting to envision the organization and sustainability of future societies, particularly in terms of embedding low-carbon systems into everyday social life (Bhatasara 2015).

Beck views current environmental challenges as elements of a "risk society," a new social formation that has replaced earlier industrial societies (Bhatasara 2015). Beck (2010) addresses the intersection of social inequality and climate change, arguing that climate change both globalizes and intensifies existing inequalities. He emphasizes the importance of the concept of social vulnerability in climate change debates, noting that it is central to understanding the potentially devastating consequences of climate change. As an adaptive response, Beck advocates for what he terms local sustainability.

Theories on the relationship between structure and agency are widely employed in environmental studies but have been rarely applied to climate research. Positioned between extreme individualism—which explains social phenomena through individual actions—and structuralism—which interprets them as outcomes of broader social and economic systems—these theories highlight the interplay between individuals and the structures in which they are embedded (Bhatasara 2015).

Across the globe, scholars have applied Bourdieu's concepts of *habitus*, *field*, and *capital* in climate change research, particularly in relation to adaptation. Giddens's structuration theory has also been used in this context. Bhatasara, however, offers a critical perspective, arguing that the theory places too much emphasis on agency while neglecting to define concepts necessary for scientifically identifying and representing structures as external, causal, and real.

The notion of social capital has been widely utilized, as seen in the works of Pelling (1998), Adger (2001, 2003), and Pelling and High (2005), though often without explicit reference to Bourdieu. Sysak (2013), for example, employed Bourdieu's theory of practice to show how power structures support resilience in northern Victorian communities in Australia. Similarly, Riera and Pereira (2009) drew on the concepts of *habitus*, *field*, and *capital* to examine how farmers in the Argentine Pampas view climate risks and adopt irrigation as an adaptation strategy. Their study highlighted that *habitus* shapes farmers' interpretation of climatic events as threats, while their challenges are tied to symbolic and other forms of field-specific capital (Bhatasara 2015, p. 223).

Other researchers have also applied Bourdieu's ideas in diverse contexts. De Bruijn and van Dijk (2005a, b) used the concept of *habitus* to explain the cultural resources farmers in Mali rely on when making decisions under climate variability. Lumosi (2014) applied Bourdieu's theory to investigate adaptation strategies and the surrounding social environment in Kenya. Likewise, Gukurume (2013) emphasized the role of social capital as a vital adaptation mechanism in his study of the Bikita district in Zimbabwe (Bhatasara 2015, p. 224).

Gutsa (2010)'s findings align with Giddens's structuration theory by highlighting the dual significance of both structure and agency. His study shows that elderly people in Chivi draw on their knowledge and social networks to confront the interconnected challenges of climate change and HIV/AIDS, which threaten their livelihoods (Bhatasara 2015, p. 224). However, Giddens's work is only sparsely referenced, resulting in a lack of deeper theoretical grounding for explaining local-level adaptation in a way that is sociologically robust and clearly articulated (Bhatasara 2015, p. 224).

The sociology of science and technology addresses the specific challenges faced by researchers in studying the relationship between nature and society. From a human ecology perspective, sociology contributes to climate change research by emphasizing the socio-spatial dimensions of climate change and the complex interactions between people and their environments across various spatial and temporal scales. Sociological inquiries into climate change can be grouped into areas such as the sociology of sustainable consumption, science and technology studies, the sociology of education, and feminist sociology. Advancing this field requires innovative theoretical contributions, greater acknowledgement of the dialectical relationship between society and nature, and the promotion of inter- and transdisciplinary collaboration (Bhatasara 2015, p. 230).

Critical sociologists provide analyses that highlight the political aspects of climate change and raise crucial questions frequently neglected in climate change studies. Sociology comprises various fundamental concepts that enable the development of key questions, such as power, ideology, mediation, capital, colonialism, oppression, and society. Critical techniques, such as Jacques Derrida's deconstruction, Foucault's discourse analysis, and post-structuralism, are vital for undertaking climate change research (Bhatasara 2015).

5.10 Anthropology for Climate Change Research

Barnes et al. (2013) identify three fundamental approaches to anthropological research: ethnographic insights, historical perspective, and a holistic vision, which are essential for modern understandings of climate change. An ethnographic study facilitates intimate observation of individuals and the identification of climate change implications within communities. Historical study facilitates the interpretation of meteorological occurrences via folklore and art, while also informing agricultural and health activities in response to these phenomena. The anthropological holistic perspective facilitates climate study. Anthropological study elucidates the social, cultural, and political processes that influence climate disputes (Barnes et al. 2013).

5.11 Interdisciplinarity of Climate Change Research

Islam and Kieu (2021) articulated four fundamental inquiries: the causes of climate change, the characteristics of its effects on humanity, the factors contributing to our insufficient reactions so far, and the prospective tactics for effective intervention. The paper primarily investigates the interdisciplinary approach of sociology in analyzing climate change. The paper advocates for interdisciplinary collaboration to understand and address the complex problem of climate change. The complex problem of climate change cannot be understood or addressed without exposing its deep linkages to the socio-political issues in our societies (Islam and Kieu 2021).

Sociological research on climate change is utilizing an inter—trans-disciplinary methodology in its studies. In the field of climate research within sociology, it is crucial to establish a novel sociological framework and seek paradigm shifts in the discipline. It is also pursuing a paradigm shift in sociological approach. Sociologists must promote transdisciplinary collaboration and multidisciplinary cooperation in climate change research. Researchers advocate for the advancement of multi-method frameworks among sociologists, encompassing qualitative sociological research that integrates GIS or remotely sensed data; investigations that analyze ecological data through the prism of power dynamics, and the expansion of the notion of "landscape" to encompass political, cultural, and social dimensions in addition to geological or biological factors. Enhance support for networks and collaborations with natural scientists and engineers; augment training for sociologists in natural science research methodologies for present and future sociologists involved in global climate change research. A sociological analysis of climate change must include the wider community of academics, policy specialists, and officials engaged in climate change research and agenda formulation (Nagel et al. 2010).

Lockie (2013) examines the convergence of ecology and sociology, especially within multidisciplinary and transdisciplinary fields such as sustainability science (Bhatasara 2015, p. 224). Koehrsen et al. (2020) assert that while many sociologists engage in Climate Change research, their results primarily manifest in inter- and trans-disciplinary publications.

5.12 Conclusion

Research on climate change has been conducted predominantly by environmental scientists and international bodies such as the Intergovernmental Panel on Climate Change (IPCC). While the human dimensions of climate change are often examined within the domains of economics and psychology, the potential contributions of sociological and anthropological perspectives remain underexplored. Scholars increasingly emphasize that climate change is not only an environmental concern but also a profoundly social and political issue, closely tied to questions of inequality and justice. At the global level, for instance, the voices of indigenous communities are frequently marginalized in climate discourse and policy processes. In this context, the role of eco-social workers becomes particularly significant, as their grassroots interventions can have a substantial impact. Drawing on systemic and structural approaches, social work practice benefits greatly from integrating sociological, anthropological, and broader interdisciplinary perspectives.

References

Adger WN (2001) Social capital and climate change (Tyndall Centre for Climate Change Research Working Paper No. 8)

Adger WN (2003) Social capital, collective action and adaptation to climate change. Econ Geogr, 79(4):387–404

Barnes J, Dove M, Lahsen M, Mathews A, McElwee P, McIntosh R, Moore F, O'Reilly J, Orlove B, Puri R, Weiss H (2013) Contribution of anthropology to the study of climate change. Nat Clim Chang 3(6):541–544

Beck U (2010) Climate for change or how to create a green modernity? Theory, Cult Soc, 27(2–3):254–266.
Bhatasara S (2015) Debating sociology and climate change. J Integr Environ Sci 12(3):217–233. https://doi.org/10.1080/1943815X.2015.1108342
Boykoff MT (2008) Lost in translation? United States television news coverage of anthropogenic climate change, 1995–2004. Clim Chang 86(1):1–11. https://doi.org/10.1007/s10584-007-9299-3
Brechin SR (2008) Ostriches and change: A response to 'global warming and sociology'. Curr Sociol, 56(3):467–474.
Brulle RJ (2015) Sociological theory after the end of nature. In: Scott R, Kosslyn S (eds) Emerging trends in the social and behavioral sciences: an interdisciplinary, searchable, and linkable resource. Wiley, pp 1–14. https://doi.org/10.1002/9781118900772.etrds0323
Brulle RJ, Carmichael J, Jenkins JC (2012) Shifting public opinion on climate change: an empirical assessment of factors influencing concern over climate change in the US, 2002–2010. Clim Chang 114(2):169–188. https://doi.org/10.1007/s10584-012-0403-y
Brundtland GH (1987) Report of the World Commission on Environment and Development: Our Common Future. https://sustainabledevelopment.un.org/content/documents/5987our-common-future.pdf
Chitale V, Behera MD (2014) India's biodiversity hotspots face climate change challenges. Nature India. https://doi.org/10.1038/nindia.2014.21
de Bruijn M, van Dijk H (2005a) Introduction: Climate and society in Central and South Mali. In M de Bruijn, H van Dijk, M Kaag K van Till (Eds.),Sahelian pathways: Climate and society in Central and South Mali (African Studies Centre Research Report 78, pp 1–15)
de Bruijn M, van Dijk H (2005b) Moving people: Pathways of Fulbe pastoralists in the Hayres. In M de Bruijn, H van Dijk, M Kaag, K van Till (Eds.),Sahelian pathways: Climate and society in Central and South Mali (African Studies Centre Research Report 78, pp. 247–287).
Declaration of Indigenous Peoples on Climate Change (2002) Australian Indigenous Law. Reporter 7(1):97–101
Delanty G, Strydom P (2003) Philosophies of social science: the classic and contemporary readings. Open University Press
Dunlap RE, Brulle RJ (2015) Climate change and society: a sociological perspective. Oxford University Press, Oxford
Giddens A (2009) The politics of climate change. Polity Press
Grundmann R, Stehr N (2010) Climate change: what role for sociology? A response to constance lever-tracy. Curr Sociol 58(6):897–910. https://doi.org/10.1177/0011392110372720
Grundmann R, Rhomberg M, Stehr N (2012) Nature, climate change and culture of social science. In: Almlund P, Jespersen PH, Riis S (eds) Rethinking climate change research: clean technology, culture and communication. Ashgate Publishing Ltd, pp 133–142
Gukurume S (2013) Climate change, variability and sustainable agriculture in Zimbabwe's rural communities. Russ J Agric Socio, 2(2):89–100
Gutsa I (2010) An ethnography of the impact of climate change, HIV and AIDS on the elderly's access to social services: the case of Chivi district. Paper presented for the2nd International Conference: Climate, Sustainability and Development in Semi-arid Regions; Fortaleza Ceara, Brazi
Harlan SL, Pellow DN, Roberts TJ, Bell SE, Holt WG, Nagel J (2015) Climate justice and inequality in R.E. Dunlap, and R. J. Brulle (eds), Climate Change andSociety: Sociological Perspectives, Oxford Academic, New York. https://doi.org/10.1093/acprof:oso/9780199356102.003.0005
Intergovernmental Panel on Climate Change (IPCC) (2023) Climate Change 2022—impacts, adaptation and vulnerability: Working Group II contribution to the Sixth Assessment Report of the Intergovernmental Panel on Climate Change, 1st edn. Cambridge University Press, Cambridge. https://doi.org/10.1017/9781009325844
Islam MS, Kieu E (2021) Sociological perspectives on climate change and society: a review. Climate 9(1):1. https://doi.org/10.3390/cli9010007
Klein N (2014) This changes everything: capitalism vs. the climate. Simon & Schuster, New York

Koehrsen et al (2020) Climate change in sociology: Still silent or resonating? Curr Sociol, 68(6), 738–760.https://doi.org/10.1177/0011392120902223
Lever-Tracy C (2010) Sociology still lagging on climate change. Sociol Res Online, 4(1):1–15
Lidskog R, Mol APJ, Oosterveer P (2014) Towards a global environmental sociology? Legacies, trends and future directions. Curr Sociol,63(3):336–339
Lockie S (2013) Sociology and global environmental change. In International Social Science Council UNESCO (Eds.), World social science report:Chang global environment, UNESCO OECD, pp. 509–511
Lumosi CK (2014) How community practitioners adapt to climate change: Integrating climate information in adaptation practices: Case studies from Kenya(Unpublished master's thesis). Wageningen University.
Magni G (2017) Indigenous knowledge and implications for the sustainable development agenda. Eur J Educ 52(4):437–447
Mulligan M (2014) Sustainability models, concepts and principles: An introduction to sustainability. Taylor Francis
Nagel et al (Eds.) (2010) Workshop on sociological perspectives on global climate change. Washington, DC: Natl Sci Found
Norgaard KM (2018) The sociological imagination in a time of climate change. Glob Planet Chang 163:171–176. https://doi.org/10.1016/j.gloplacha.2017.09.018
Pelling M (1998) Participation, social capital and vulnerability to urban flooding in Guyana. J Int Dev, 10(4):469–486.
Pelling M, High C (2005) Understanding adaptation: What can social capital offer assessments of adaptive capacity? Glob Environ Chang,15(4):308–319
Riera C, Pereira SG (2009) Irrigated agricultural production and adaptation to climate change in the Argentinean Pampas: An analysis from a sociotheoreticalperspective. Int J Sustain Stud, 1(1), 35–39.
Sustainable Development Goals Report (2022) https://unstats.un.org/sdgs/report/2022/
Sunderlin WD (1995) Global environmental change, sociology, and paradigm isolation. Glob Environ Chang, 5(3):211–220
Sysak TS (2013) Drought power and change: Using Bourdieu to explore resilience and networks in two Northern Victoria farming communities(Unpublished PhD thesis). University of Melbourne, Dep Resour Manag Geogr
United Nations & World Bank Group (2018) Making every drop count: an agenda for water action. United Nations. https://www.unwater.org/publications/making-every-drop-count-an-agenda-for-water-action
United Nations Environment Programme Finance Initiative (2023) Sectoral risk briefing: climate risks in the agriculture sector. United Nations Environment Programme. https://www.unepfi.org/wordpress/wp-content/uploads/2023/03/Agriculture-Sector-Risks-Briefing.pdf
Walker N (2011) Sociology and climate change. In: Dryzek JS, Norgaard RB, Schlosberg D (eds) The Oxford handbook of climate change and society. Oxford University Press, pp 77–93
UNDP (2023) The Climate Dictionary: Speak climate fluently. UNDP
Urry J (2009) Sociology and climate change. Sociol Rev, 57(2_suppl), 84–100. https://doi.org/10.1111/j.1467-954X.2010.01887.x
Zalasiewicz et al (2008) Are we now living in the Anthropocene? GSA Today. Geol Soc Am, 18(2):4–8

Chapter 6
Community-Centric Flood Management: Insights from a Citizen-Led Case Study

Vinumol Devassy, M. Riswan, and M. Pushpam

Abstract Floods remain one of the most destructive natural hazards globally, causing extensive damage to lives, property, and ecosystems, particularly in vulnerable river basins. The state of Kerala, in southern India, has witnessed recurrent and severe flood events in recent years, prompting renewed interest in innovative, participatory approaches for disaster risk reduction. Against this backdrop, the increasing integration of community-led action, technology, and real-time information sharing represents an important shift in flood management approaches. Building on these trends, the following case study examines a community-led initiative for flood management along the Meenachil River in Kerala, India, highlighting the integration of weather warning systems through a series of community-monitored rain and river gauges, weather stations and social media platforms. It explores how local communities actively engage in monitoring and disseminating critical weather information, fostering resilience against flooding. The resilience built among the community members helps in disaster risk reduction, emphasizing community engagement and technological integration for effective flood management.

Keywords Community-led flood management · Disaster risk reduction · Resilience · Technological integration · Social media communication

---

V. Devassy (✉)
Department of Sociology, Loyola College of Social Sciences (Autonomous), Thiruvananthapuram, Kerala, India

M. Riswan
Department of Sociology, South Eastern University of Sri Lanka, Oluvil, Sri Lanka
e-mail: mriswan2008@seu.ac.lk

M. Pushpam
Department of Sociology, University of Kerala, Thiruvananthapuram, Kerala, India

J. K. Joseph, A. P. Pradeepkumar (eds.), *Eco-Social Work for Disaster Risk Reduction*, Disaster Resilience and Green Growth, https://doi.org/10.1007/978-981-95-4885-9_6

6.1 Introduction

Floods are the most destructive natural disaster in the world. Every year, this natural disaster disturbs millions of people's lives. There are several factors contributing to the occurrence of flood situations. They are heavy rainfall, sea overflow, rapid melting of snow, dam failures and breakage (Lai 2023). The overflow of water causes significant changes in the environment and human settlements. Additionally, it impacts the activities of individuals and communities. Therefore, it is crucial to develop and implement strategies to reduce flood risks. Flood control methods are a type of flood management method that involves various methods aimed at reducing the effects of floodwater.

Flood management refers to the strategies and practices that are used to minimize the negative impact of flooding on people, property and the environment. According to Benson et al. (2016), flood management is a collective term for a series of activities following sustainable development principles, aiming to balance human needs with nature while normalizing flood control efforts. Effective flood management requires planning, preventive actions, and response measures that handle water overflow situations (Fiveable 2024). It focuses on strategies that protect lives, properties, and natural habitats. However, addressing strategies for flooding is a complex task. It encompasses a range of environmental, technical, economic, and social dimensions. Effective flood management requires clear communication and coordinated action among multiple stakeholders, including government agencies, non-governmental organizations, local communities and international organizations (Rao and Franklin 2024).

Traditional flood management relied on two main approaches, including structural (keep floodwater away from people) and non-structural (keep people away from floodwater) measures to reduce the adverse consequences of a flood event. These management methods aim to mitigate, reduce or eliminate impacts before the flood occurs (Wang et al. 2022). Communities can play a key role in the non-structural management strategies. People are no longer seen as recipients; rather, they have become critical stakeholders who have actively contributed to the management of community flood reduction initiatives (World Meteorological Organization 2017).

The community-centric approaches are founded on the understanding that floods are not only viewed as a hydrological event but also as complex socio-ecological phenomena. The impact of flooding is different across communities and individuals. It is influenced by factors including socio-economic status and housing quality and social network and access to resources. In this context, effective mitigation requires interventions that are not only technically sound but also socially equitable and locally adaptable (Sustainability Directory 2025). Community involvement is more effective when people are fully informed, empowered and trained with the necessary skills. Therefore, government and public authorities play a crucial role in facilitating and supporting community-based flood management initiatives, like engaging the community throughout the whole phase of flood management activities, from assessment, design, implementation, to monitoring and evaluation. It is ensuring that adopted measures remain equitable and effective for the entire affected population (Oluseyi et al. 2011). Consequently, community participation becomes an essential element at every stage of flood management, including disaster prevention and preparedness, emergency response to and recovery from flood disasters.

Flooding is one of the most frequent natural disasters affecting numerous states in India. Out of the total geographical area of 329 million hectares (mha), more than 40 million hectares are considered flood-prone. In recent years, India has witnessed an upward trend in flood-related damages (Kumar 2018). Against this backdrop, Kerala experiences the highest rainfall during the monsoon season. It has witnessed repeated flooding situations, particularly in recent decades. It impacts local communities, agriculture, and infrastructure. Among the rivers in Kerala, the Meenachil River flows through the Kottayam district. Its basin has been identified as particularly vulnerable to flooding. This situation arises from rapid urbanization, inadequate drainage systems and the effects of climate change (Sukumaran and Birkinshaw 2024). Seasonal floods in the river basin disrupt agriculture, damage infrastructure, displace families and create long-term vulnerabilities across both rural and urban settings.

In India many flood management strategies have largely relied on top-down approaches emphasizing structural interventions. This approach provides limited relief; it often fails to address the needs and problems of affected people. This gap has highlighted the recognition and importance of the community-centric flood management approaches, which integrate local participation and knowledge. In this regard, the Meenachil river basin presents a compelling case for examining such community-driven initiatives. The programs implemented in the Meenachil River basin top the list due to the significant role of public participation, which has been described as a "silent revolution" transforming local flood resilience efforts in the region (Devassy 2025). In this sense this article explores how local communities take part in the flood management activities. It aims to highlight disaster risk reduction methods by emphasizing people's participation.

Chapagain et al. (2024) present the development, validation, and application of the Flood Resilience Measurement for Communities (FRMC) approach, designed by the Zurich Flood Resilience Alliance to assess and enhance community flood resilience globally. The FRMC framework integrates a comprehensive set of 44 resilience sources, structured within five capitals—financial, human, natural, physical, and social—and aligns with disaster risk management cycles. It employs quantitative tools such as principal components analysis and reliability testing to ensure validity and consistency, demonstrating high reliability across diverse communities. The approach emphasizes participatory data collection involving local stakeholders to capture community perspectives and tailor resilience-building efforts effectively. By analysing data from over 13,000 community assessments across 20 countries, the study establishes a taxonomy of community resilience profiles, enabling comparisons and targeted interventions. Overall, the FRMC provides a standardized, systems-oriented methodology that supports policymakers, practitioners, and communities in measuring, understanding, and strengthening flood resilience in a holistic and context-sensitive manner.

6.2 Community-Centric Approach

Since flood management is a socio-ecological issue rather than a hydrological or engineering problem, it certainly demands the involvement of the people in the region. In the report of the Intergovernmental Panel on Climate Change (2022), the present world requires public participation in matters concerning socio-ecological systems. Over the past decades, the Community-Based Early Warning System (CBEWS) and Community-Centered Disaster Risk Reduction (CBDRR) have shifted the emphasis from top-down warnings to locally produced knowledge, monitoring and action, thereby improving lead time for response, enhancing local capacities and reducing vulnerability in situations where formal systems fail to reach or address. Empirical and review studies say that locally driven monitoring, interpretation of environmental signals and community dissemination channels strengthen the four pillars of early warning (Risk Knowledge, Monitoring, Dissemination, and Response) and help convert warnings into timely action (Macherera and Chimbari 2016; Josè Moisès D and Nyam, 2025)

The importance of community engagement in flood management is now recognised in both international and national policy frameworks. The Sendai Framework for Disaster Risk Reduction (2015–2030) highlights the need for "people-centred multi-hazard early warning systems" that are accessible, understandable, and trusted by local populations (UNDRR 2015). Case studies compiled in UNFCCC and UNDP reports similarly demonstrate that locally managed flood early warning systems are more cost-effective, sustainable, and inclusive than top-down alternatives (UNFCCC 2019).

India's experience underscores the significance of this shift. In India, the National Disaster Management Authority (NDMA) has issued guidelines promoting community-based disaster risk reduction (CBDRR), emphasising the role of local knowledge and volunteer networks (NDMA 2019). Kerala, during the 2018 flood, had witnessed community engagement in post-flood management. The importance of a bottom-up approach in disaster management, comprising "Community-Based Disaster Risk Management" (CBDRM), which can transform vulnerable communities into disaster-resilient communities, is emphasized in the study of Joseph et al. (2020).

Having experienced the harmful effects of natural hazards like floods, landslides, and drought, the Kerala community is proactively aligning itself towards the Community-Based Disaster Risk Reduction approach. This paper explores the Community-driven Citizen Science Network in the Kottayam district of Kerala, India—the "Meenachil River-Rain Monitoring Network" (MRRM) spearheaded by a community-driven people's forum named "Meenachil River Protection Council." This council has a long history of more than three decades of involvement in the management of the Meenachil River Basin in the Kottayam district of Kerala, India.

6.3 Meenachil River in the Kottayam District of Kerala, India

Situated in central Kerala, the Kottayam district has a unique terrain, bordered by the majestic Western Ghats on the east, the Vembanad backwaters on the west, and the paddy fields of Kuttanad. The district of Kottayam covers an area of 2208 square kilometres. The district is longitudinally divided into three geographic regions: *highland, midland, and lowland*, with the midland region being the major one. Meenachil River is 78 km long and flows through the Kottayam district, after originating from Arikunnumudi (1097 m above mean sea level) of the Western Ghats. It lies between 9° 25′ N to 9° 55′ N latitudes and 76° 22′ E to 76° 57′ E longitudes and spreads over a catchment area of 1272 sq.km. The drainage pattern observed in the basin is dendritic and sub-dendritic (Vincy et al. 2015).

The Meenachil River Basin is integral to the local geography and ecology. It is the primary river that caters to the needs of the region. After originating from the Western Ghats it meanders through the basin, providing water for irrigation and supporting local systems. The river supports agriculture in the region. By hosting the vast flora and fauna in the region and being the habitat for multiple species, the Meenachil River offers its ecosystem services to the area. The people in the region have cultural ties to the river; they have several practices which are associated with this river and it has social and cultural significance in the area. The economic importance of the river cannot be undermined as it supports the livelihoods in many in the basin and promotes water tourism, etc.

6.4 Meenachil River Protection Council (MRPC)

The inception of the Meenachil River Protection Council (MRPC), a people's forum in the region, dates back to the late 1980s when a task force (by the people) emerged in response to the proposed Adukkam Dam at the Teekoy Grama Panchayat in the Kottayam district by the Government of Kerala. Concerns about the potential ecological harm prompted community opposition to the dam. This resistance was rooted in a deep commitment to safeguarding the Meenachil River in the years to come. Initially, the activities of the MRPC focused on water availability, unscientific dam construction, soil mining, and pollution in the river basin. Later, after the silver jubilee of its formation, the MRPC took several measures to expand its activities. Alongside the region faced issues like heavy rain, flash floods, and landslides with an increased frequency than expected; prompting an increased emphasis on disaster management activities. Major domains of the activities of the MRPC are depicted in Fig. 6.1.

Figure 6.1 illustrates the various domains in which the activities of the MRPC are concentrated. The people's forum engages in activities that contribute to the protection of the Meenachil River Basin, natural hazard mitigation and also in increasing

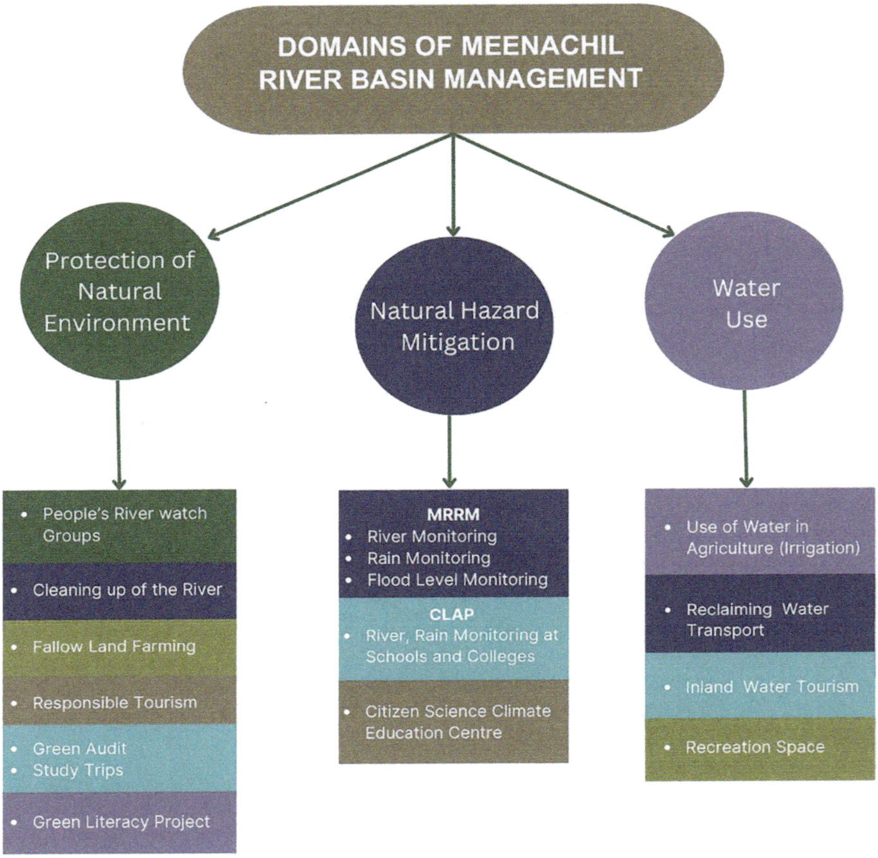

Fig. 6.1 Domains of Meenachil River Basin Management by Meenachil River Protection Council

the water use in the Meenachil River Basin. Under these three main domains, the MRPC offers multiple decentralized platforms that enhance activism, volunteerism, collaboration, education and evaluation within the region. Collectively, these participatory practices reinforce the community oriented character of this people's forum. Meenachil River Rain Monitoring Network (MRRM) is one among the significant initiatives which is ongoing in the Meenachil River Basin Management, which has a strong volunteer base.

6.5 Meenachil River--Rain Monitoring Network (MRRM)

The river–rain monitoring activity was initiated through a WhatsApp group in the aftermath of the catastrophic floods of 2018 and gradually evolved into the Meenachil River–Rain Monitoring (MRRM) Network by 2021. In 2020, based on the group's systematic observations and timely warnings regarding rainfall and rising water levels in the Meenachil River, approximately two hundred residents of the Koshammattom Colony in Kottayam district, Kerala, were successfully relocated. This proactive intervention proved critical in preventing the loss of lives during an unexpected flash flood event.

The effectiveness of this initiative drew recognition from both the local community and the district administration. Its success was rooted in the indigenous knowledge of the local population, particularly their experiential understanding of the relationship between water levels in the upland regions of the Meenachil River Basin and the subsequent flooding of the midland areas within a predictable timeframe. This traditional ecological knowledge, when combined with systematic observation, played a decisive role in safeguarding hundreds of lives.

Acknowledging the importance of generating reliable hydrological data for the local communities, the core team of the Meenachil River Protection Council subsequently initiated the installation of rain gauges in strategically identified households, thereby institutionalizing rain and water level monitoring practices. The initiative is being supported by climate scientists from the Indian Institute of Tropical Meteorology (IITM), Pune, the Department of Environmental Sciences, University of Kerala, Saintgits Engineering College, Pathamuttom, Koll Science Foundation, and Thopps Technologies. At present, the MRRM has gained national visibility and functions as a community-based early warning system. It integrates citizen-led observation networks with scientific methods of data recording and analysis, facilitated by trained volunteers, thereby establishing a model of collaborative disaster preparedness and environmental stewardship.

The Meenachil River Resource Management (MRRM) represents a collaborative initiative that brings together a wide network of stakeholders, including hundreds of rain gauge observers, over a dozen river gauge observers, numerous live data providers, College Climate Action Groups, and Gram Panchayats. As a member of the Inter-Agency Group (IAG) under the District Disaster Management Authority, the Meenachil River Protection Council plays a pivotal role in leading this citizen science initiative. The program is designed to strengthen disaster preparedness, ensure year-round watershed protection, prevent drought, and facilitate systematic data collection, research, analysis, and information dissemination.

6.5.1 *Working of MRRM*

The MRPC progressively expanded its activities by distributing rain gauges to individual volunteers and subsequently securing institutional support from local self-government bodies. Through sustained petitions and proposals submitted by MRPC volunteers, financial assistance was eventually sanctioned, despite initial reluctance on the part of panchayats to support the initiative. With this approval, rain gauges were installed across nearly six panchayats, ensuring ward-level coverage through strategically identified locations and designated volunteers. Panchayat members also played an active role by nominating volunteers from their respective wards, thereby reinforcing the collective responsibility for the initiative.

As part of the monitoring mechanism, a WhatsApp group was established for each ward, where volunteers were instructed to report daily rainfall readings at 8:30 a.m. These groups included the elected ward member, a representative from the MRPC, and a designated Team Lead among the volunteers who oversaw compliance and encouraged timely reporting. This structure not only ensured accuracy and accountability but also fostered a participatory model of citizen science.

The growing significance of the data attracted collaborations with scientific and academic institutions. The Koll Science Foundation, in particular, provided a coordinator responsible for compiling ward-level rainfall data and reporting it to the central *Save Meenachilar* WhatsApp group. Volunteers were formally recognized through the issuance of identification cards, enhancing their visibility and legitimacy in the project.

Initially, the MRPC's activities were concentrated in the highland regions of Kottayam district, where the population predominantly comprises farmers and daily wage labourers. Over time, the benefits of the initiative extended to the midland regions as well, where improved rainfall monitoring provided critical insights for both disaster preparedness and agricultural planning. Presently, rain gauges have been systematically installed across multiple Grama Panchayats, including Poonjar-Thekkekkara, Theekoy, Thalanad, Thalappalam, and Meenachil, among others, thereby establishing a decentralized, community-based rainfall monitoring network that integrates grassroots participation with institutional and scientific collaboration. In the words of a 37 year-old volunteer from Thalappalam Panchayat,

> I found the rain monitoring initiative of MRPC to be very efficient. Information shared by the MRPC during the rainy season was very useful. I found this interesting, and thus, I became a volunteer in assessing the rain on a day-to-day basis. I would continue to do this. This helps our region in multiple ways.

For the volunteers and team leaders of the Meenachil River–Rain Monitoring (MRRM) Network, the practice of daily reporting has evolved into an engaging and meaningful routine. A typical volunteer records observations manually, creating tables with columns and rows on paper to enter rainfall data from the installed gauges. As a team leader, the volunteer not only documents personal observations but also collects and consolidates the data reported by other volunteers within the ward. This compiled information is then shared in the designated WhatsApp group for team leaders, where it is further organized and summarized for centralized analysis.

Through this sustained engagement, volunteers gradually develop observational skills and localized insights. They begin to recognize patterns, such as the likelihood of rainfall occurring simultaneously across multiple areas, as well as the varying intensity of precipitation. This growing capacity for interpretation highlights how grassroots participation fosters both community-based scientific literacy and proactive disaster preparedness. He said,

> During the earlier days of my childhood, I heard the predictions of my ancestors concerning rain and weather conditions by observing nature. Later I depended on government forecasts regarding the various weather alerts in the region. However, now I could predict the weather updates in the region and all predictions now have a scientific base, so I found this interesting and useful.

The Meenachil River Protection Council has installed over 200 rain gauges across diverse locations to measure spatial variations in rainfall. In addition, 13 river gauges have been established at strategic points along the river to monitor fluctuations in water levels, particularly flood conditions. Instruments for recording temperature have also been installed at selected sites, thereby expanding the scope of the monitoring system beyond precipitation and river flow.

The installation of these instruments has been facilitated through collaborations with the Centre for Water Resources Development and Management (CWRDM), Calicut, and the Indian Institute of Tropical Meteorology (IITM), Pune. Further, the Indian Institute of Architects (Cochin chapter) sanctioned financial assistance of ₹135,000 for the procurement of additional rain gauges and thermometers, following a request made during a lecture delivered by Dr. Roxy Mathew Koll of IITM. This support has enabled the acquisition of higher-quality instruments, which are now being systematically deployed across different watersheds rather than concentrated in a limited set of locations.

At present, the Meenachil River–Rain Monitoring (MRRM) Network represents a pioneering model, as no comparable citizen-led system exists for any of the other 44 rivers in Kerala that provides real-time scientific information through community participation. The initiative has recently gained state-level recognition, receiving special mention in the Kerala State Action Plan for Climate Change (2022–23), a development that has instilled a sense of pride among the volunteers and further validated the significance of their contributions.

> "The Meenachil River Rain Monitoring Programme, a citizen science network equipped with rain gauges and water scales across the Meenachil River in Kottayam that monitors rainfall and river water level, is now assisting the District Administration in taking necessary actions during extreme weather events". (Kerala State Action Plan on Climate Change 2023–2030, 2022)

All these initiatives continue to function effectively owing to the sustained commitment and active participation of the local community.

6.5.2 *Regarding River Monitoring*

Traditionally, community members observed river water levels and communicated them in localized terms, often by relating them to the visible length of a bridge pillar. Recognizing the limitations of such informal methods, they sought more systematic and scientifically credible measurements. This led to the creation of a "people's scale," drawn on a rock, which quickly gained popularity and wide acceptance within the community. Building on this success, the Meenachil River

Protection Council formalized these practices by developing river scales with the support of volunteers. These scales enabled residents to identify alert stages of flooding, fostering a participatory model of flood risk awareness and response.

Through these practices, citizens gradually acquired scientific knowledge about hydrological patterns and climate variability. Volunteers learned to identify rainfall intensity and distinguish patterns, moving from intuition-based observations to evidence-based classifications of rainfall as heavy, moderate, or mild. "Learning" thus became a central dimension of the initiative, empowering citizens with both experiential and scientific insights. While volunteers routinely reported data via WhatsApp groups, they were also provided with record books and pens to manually document rainfall measurements. This practice not only created reliable archives for analysis but also offered farmers—many of whom served as volunteers—an important tool for understanding precipitation trends and making informed agricultural decisions. More recently, technological tools have been increasingly integrated into the system, further enhancing the precision, accessibility, and long-term value of community-generated data.

According to Koll, climate scientist, IITM, Pune.

> In this era of climate change, the local variations in the climate are so significant. Lacking appropriate data is always challenging for the scientific and academic community. In this regard, what these MRRM (Meenachil River-Rain Monitoring Network) Volunteers are doing is highly appreciable.

The data generated through this initiative provides valuable inputs for the scientific community, particularly in studying micro-level variations in local climate. Beyond its academic relevance, the information also serves as a critical resource for disaster management, enabling timely interventions and preparedness. Local residents, through their active engagement, have developed an enhanced awareness of climatic variability and rainfall intensity in their respective areas. Recognizing the significance of such citizen-led monitoring, CWRDM has been providing training to volunteers to strengthen their observational and recording capacities. Similarly, the Kerala State Disaster Management Authority (KSDMA) has partnered with the Meenachil River Protection Council by supporting the installation of river and rain gauges in select locations. These institutional collaborations have not only broadened the scope of the initiative but also fostered more responsible leadership and greater volunteer participation. As a result, community-driven disaster preparedness and mitigation measures have been significantly strengthened. The active involvement of people equipped with accurate knowledge and real-time data has made managing critical situations more effective, participatory, and sustainable. According to another volunteer, from Pala,

> I am working in line with the instructions given by the main team leaders. I publish all my observations on social media in my language (Malayalam) and I found my responsibility is to share the data with the common people. The work I do can be done by anyone else. I observe the rain gauge, then look at the river gauge and notice the speed of the flowing water. After months of experience, I can predict if a particular amount of rain has been received, this much water will be seen in the river and the like.

This citizen science network *integrates technical expertise and local knowledge* in the best possible ways. The central inference to be noted here is that the layman generates, records, and the data generated through this initiative is systematically analysed, marking a significant shift in knowledge production. It is particularly noteworthy that the traditional monopoly of administrative bodies and government officials over hydrological data has been disrupted. Instead, local residents—ordinarily excluded from such processes—have assumed the role of data creators and informants, reporting on rainfall intensity and river conditions. This democratization of knowledge has facilitated more effective communication between community members, local authorities, and government officials, especially in issuing and understanding alert messages related to river conditions.

The monitoring and maintenance of river gauges are managed through the decentralized platforms of the Meenachil River Protection Council. These include the CLAP (Climate Leadership in Action Program) units established in schools and the *Kaavalmaadangal* (People's Watch Groups of the Meenachil River), which oversee and report irregularities or threats in and around the river ecosystem. In addition, the initiative emphasizes capacity-building by providing training to ordinary citizens in specific technical skills, thereby enhancing community resilience and ensuring that the benefits of knowledge and practice are reciprocally shared.

6.6 Discussion and Analysis

The case demonstrates that effective flood management can be achieved through the coordinated efforts of the community. An analysis of this community-centric, citizen-led flood management model highlights three interrelated dimensions—technical, institutional, and socio-cultural (Alexander 2008; Bollin et al. 2006). Each of these dimensions contributes significantly to strengthening local flood resilience by integrating community participation, digital communication, indigenous knowledge exchange, and multi-stakeholder partnerships.

Figure 6.2 illustrates the three dimensions—technical, institutional, and socio-cultural—of the community-based flood management model in the region. The technical dimension, traditionally the responsibility of government officials, is now effectively managed by local residents, reflecting the utilization of indigenous knowledge and experiential expertise. The institutional dimension emerges from the formal recognition of volunteers, instilling a sense of pride and reinforcing their role as authentic contributors to both community welfare and personal development. The socio-cultural dimension emphasizes the social cohesion and solidarity fostered through collective action. Additionally, by involving and educating the next generation, the initiative ensures the sustainability of the program, embedding knowledge and practices that support long-term resilience and community engagement.

Fig. 6.2 Elements in a community-centric flood management

6.6.1 Technical Dimension

6.6.1.1 Data Collection and Local Surveillance

In the domain of disaster risk reduction, the state has traditionally borne the responsibility of collecting field-level data and issuing graded alerts—such as red, orange, or yellow warnings—for communities residing in river basins. However, the absence of reliable, fine-grained data from localities frequently undermines the timeliness and effectiveness of governmental interventions. This challenge is compounded by a shortage of field personnel, which constrains the capacity of state agencies to access remote or highly vulnerable regions.

Emerging evidence underscores the growing significance of citizen science and participatory monitoring as vital complements to state mechanisms, highlighting the value of a bottom-up approach in climate adaptation strategies (Bonney et al. 2009; Irwin 1995). The systematic installation of household-level rain gauges and coordinated regional monitoring initiatives have enabled communities to generate highly localized climatic datasets. These practices not only strengthen the technical capacity of grassroots actors but also foster empowerment and ownership over disaster preparedness processes (Danielsen et al. 2014).

Importantly, the credibility of these citizen-generated datasets is reflected in their increasing demand by NGOs and research institutions, which employ them to document micro-level climatic variations and strengthen adaptive planning (Buytaert et al. 2014). A particularly noteworthy aspect is the dedication of individuals who regularly monitor rainfall and river-gauge levels to establish correlations between precipitation and flood risks. What frequently begins as the initiative of a single household member often evolves into collective family participation, thereby reinforcing resilience and embedding a culture of community-based environmental stewardship.

6.6.1.2 Use of Social Media and Other Digital Platforms

The use of social media and digital platforms has become integral to community-based disaster communication and data management. Among local residents, WhatsApp groups serve as the primary channel for information exchange, enabling individual community members to transmit recorded data in real time. This information is subsequently streamlined through *Bhoomika*, a data collection centre established by the MRPC, where a local resident is employed to systematize and verify the inputs. The curated datasets are then uploaded to the MRPC's official website and disseminated further through social media platforms such as Facebook, thereby reaching a wider public and alerting community members to emergent risks.

This model represents one of the first documented regional initiatives for integrating digital platforms into the systematic monitoring of rainfall and river-gauge data. Importantly, the process illustrates how digital infrastructures and social media can enhance citizen science practices by facilitating rapid information sharing, fostering transparency, and bridging the gap between local data production and broader institutional use (Houston et al. 2015; Reuter and Kaufhold 2018). Moreover, the integration of WhatsApp-based networks with formalized data repositories reflects the increasing recognition of Information and Communication Technologies (ICTs) as vital tools in community-driven environmental monitoring and disaster preparedness (Meier 2015; Palen and Anderson 2016).

6.6.1.3 Monitoring and Infrastructure Maintenance

Community-led monitoring has increasingly become a cornerstone of localized disaster resilience. In the studied region, continuous community-based observation is coupled with the routine maintenance of critical flood control infrastructure. Such practices translate local ecological knowledge into concrete conservation efforts, thereby strengthening both community and environmental resilience (Agarwal 2001; Berkes 2009).

The maintenance of hydrological monitoring equipment—including river gauges and household rain gauges—is also coordinated by community volunteers. A member volunteer or one of the core team members would periodically visit the homes of participating volunteers to ensure functionality and accuracy of the instruments. While government officials acknowledge that maintenance and follow-up are often the most challenging aspects of disaster-related initiatives, community engagement substantially mitigates these challenges. Volunteers not only help sustain government-installed weather stations but also assume ownership of the process. This sense of collective ownership and responsibility is critical, as it encourages sustained participation and ensures the long-term viability of the initiative (Pretty 2003; Ostrom 1996).

Moreover, participation in infrastructure maintenance fosters solidarity within the community. The process is both educational and empowering, enabling residents to become active contributors to scientific and technical processes rather than passive recipients of government directives. The effective use of digital platforms for coordination further enhances community learning and cohesion. Particularly noteworthy is the involvement of senior citizens, who actively engage as "data creators", thereby challenging assumptions about technological exclusion and demonstrating intergenerational collaboration in environmental stewardship (Fernandez-Gimenez et al. 2008; Fraisl et al. 2020).

6.6.2 Institutional Dimension

6.6.2.1 Institutionalization of Community Roles

The formal recognition of community participation represents a critical step toward the institutionalization of grassroots engagement in environmental governance. Within the MRPC, community volunteers are officially registered as part of the organizational structure and are provided with badges and record books to document daily rainfall data. These entries are simultaneously communicated through WhatsApp to designated team leaders, ensuring both consistency and accountability in data management. Such formalized roles not only enhance the credibility of community contributions but also foster a sense of pride and recognition among participants. Being acknowledged by peers, neighbours, and the broader community forum reinforces the social value of their involvement, thereby motivating sustained engagement. The symbolic dimension of recognition—through badges, institutional affiliation, and acknowledgment by MRPC's core team—strengthens the volunteers' perception that their contributions are both valid and meaningful. This process reflects how community-based monitoring transitions from informal citizen action to an institutionalized practice embedded within broader governance frameworks (Ostrom 1996; Berkes 2010). Moreover, the deliberate efforts of MRPC's leadership to validate and inspire volunteers underscore the importance of institutional support in sustaining collective action. Such recognition not only legitimizes local knowledge but also situates community members as active partners in achieving broader environmental and societal goals. In this way, the institutionalization of community roles contributes to building trust, social cohesion, and long-term resilience (Pretty 2003; Pahl-Wostl 2009).

6.6.2.2 Shared Governance and Participatory Decision-Making

The activities of the people's forum under the MRPC exemplify a participatory governance model in which citizen-led initiatives attract the active involvement of bureaucrats and local authorities. Through conscious, sustained, and vigilant efforts, community members were able to hold government actors accountable, thereby establishing a system of mutual accountability. In this arrangement, the people's forum exercise significant influence without assuming the responsibilities of other governance counterparts, illustrating the potential of citizen groups to shape institutional practices. Figure 6.3 demonstrates the interdependent role of these three significant components of the MRRM functioning framework.

This dynamic fostered a tripartite collaboration among citizens, government officials, and NGOs, which enhanced both the effectiveness and efficiency of the initiative. The cultivation of mutual trust between these actors was central to its success, as it enabled the sharing of responsibilities and resources in a balanced manner. NGOs, often associated with the community side of participatory governance, further strengthened this relationship by offering technical and organizational support. Nevertheless, the continuity and effectiveness of the initiative remained rooted in the dedicated efforts of MRPC volunteers, whose persistent engagement ensured that external collaborations could be meaningfully sustained.

The MRPC's role extended to consistently briefing the district administration on the evolving conditions of the Meenachil River, reflecting their perception of this task as a civic duty. Recognition from government agencies followed in the form of technical support and institutional acknowledgment. Notably, the people's forum actively approached local self-government (LSG) institutions to advocate for the installation of rain gauges, often attempting to secure funding through climate crisis allocations. While initial attempts to persuade panchayats met with limited success, the visible effects of climate change on agricultural fields eventually validated the community's efforts. Consequently, LSGs and other authorities began recognizing the necessity of these measures and acknowledged MRPC's pioneering role.

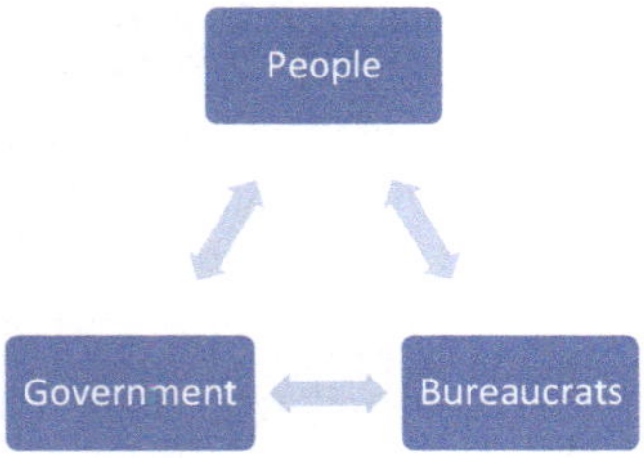

Fig. 6.3 Tripartite integrated approach in shared governance

It is important to note that the initial stages of the initiative were entirely self-financed by community members, who pooled their resources to purchase rain gauges and establish a rudimentary monitoring network. This bottom-up investment demonstrates how citizen ownership can drive institutional responsiveness, aligning with theories of participatory governance that emphasize co-production of knowledge and accountability between citizens and the state (Ansell and Gash 2008; Cornwall 2008).

6.6.2.3 Collaboration and Capacity Building

Collaborative partnerships are integral to sustaining community-led disaster risk reduction initiatives. In the case of the Meenachil River Protection Council, collaboration with external institutions—such as government departments, NGOs, and universities—emerged only after the initiative demonstrated resilience and continuity at the grassroots level. This sequencing underscores an important dynamic: the authority to engage with external actors rests with the people's forum, which retains decision-making power over when and how to form partnerships. Such autonomy reinforces community ownership and prevents premature external domination of local processes (Cornwall 2008; Ostrom 1996).

As the initiative gained credibility, diverse institutions extended support in the form of technical expertise, validation of data, and training for volunteers. Governmental departments, initially observers, began to rely on the MRPC's updates during climate emergencies, while NGOs and universities contributed by transferring knowledge and introducing technological innovations. This process reflects how participatory governance frameworks expand collaboration, embedding local voices within broader disaster management strategies and ensuring that interventions remain contextually grounded (Ansell and Gash 2008; Berkes 2009).

Capacity-building initiatives play a pivotal role in consolidating these partnerships. Volunteers receive structured training in monitoring techniques, data management, and the use of digital platforms. Schools and colleges also form part of this collaborative ecosystem, with the establishment of Climate Action Groups that monitor rainfall, river-gauge levels, and temperature. By engaging educational institutions, the initiative not only strengthens technical monitoring but also cultivates intergenerational awareness and long-term stewardship.

These collaborations highlight the complementary nature of local and institutional knowledge systems. Universities and NGOs provide scientific validation and access to technological resources, while community groups contribute place-based knowledge, continuity, and social legitimacy. The integration of these forms of expertise creates hybrid governance arrangements that enhance sustainability and embed resilience within both community and institutional structures (Pahl-Wostl 2009; Ensor and Harvey 2015).

6.6.3 Socio-Cultural Dimension

6.6.3.1 Social Learning and Knowledge Exchange

A culture of preparedness is nurtured when communities actively participate in workshops, neighbourhood education programs, and peer-to-peer learning forums. Both digital platforms and traditional outreach methods raise awareness, simplify technical processes, and foster crowdsourced innovations in flood response. Importantly, these initiatives go beyond mere technical engagement—they cultivate intergenerational learning practices. For instance, when adults in a household measure rainfall using gauges, children often observe with curiosity and enthusiasm, eventually training themselves and developing ecological sensitivity. Such involvement equips the younger generation with an early awareness of environmental changes, laying the foundation for sustainable practices (Ballard and Belsky 2010).

Farmers also benefit significantly from these practices, as real-time climate and hydrological information shared within the community reduces their dependency on external government resources. The resulting knowledge exchange enhances local adaptive capacity and creates a sense of social cohesion, whereby residents perceive themselves as growing together and collectively responding to environmental challenges. This participatory knowledge system transforms individual efforts into shared resilience, demonstrating the value of informal, community-driven education (Reed et al. 2010).

6.6.3.2 Community Ownership and Social Cohesion

The active involvement of residents as local protectors deepens their sense of responsibility toward the environment. By engaging in collective monitoring activities, communities not only safeguard themselves against climatic risks but also develop stronger social bonds. This sense of ownership is essential for sustaining long-term resilience, as it strengthens mutual trust and cooperation in times of crisis. Localized communication networks—particularly community-based WhatsApp groups—further reinforce this cohesion by ensuring rapid information exchange. These grassroots channels have proven valuable not only for households but also for decision-makers at the panchayat, district, and even state levels, thereby bridging community initiatives with formal governance structures (Aldrich and Meyer 2015; Shaw 2012).

6.6.3.3 Community-Led Education and Outreach

A distinctive feature of this socio-cultural model is its emphasis on education and outreach that emerges from within the community rather than being externally imposed. Local forums facilitate climate awareness sessions, while schools and colleges establish Climate Action Groups that integrate academic knowledge with grassroots practices. These groups are entrusted with monitoring rainfall, river gauges, and temperature variations, thereby institutionalizing environmental stewardship among youth. By embedding climate action into educational spaces, the community ensures continuity and long-term sustainability of its initiatives (Armitage et al. 2011).

Through such interlinked efforts, the socio-cultural dimension of flood resilience extends far beyond risk management. It represents a transformative process of collective learning, shared responsibility, and social empowerment that strengthens both immediate preparedness and long-term adaptive capacity.

6.7 Conclusion

When flood management processes—from data collection and surveillance to infrastructure upkeep and response coordination—are handled by the community, residents in the region evolve from passive recipients to empowered protectors. Their technical engagement, environmental stewardship, and participatory governance facilitate better, locally appropriate solutions and nurture ongoing resilience such that they become the data providers for the scientific community. The research confirms that community-led approaches produce more robust, sustainable flood mitigation, not only reducing risk but they are also building long-term adaptive capacity and agency at the grassroots level.

References

Agarwal A (2001) Common property institutions and sustainable governance of resources. World Dev 29(10):1649–1672. https://doi.org/10.1016/S0305-750X(01)00063-8

Aldrich DP, Meyer MA (2015) Social capital and community resilience. Am Behav Sci 59(2):254–269. https://doi.org/10.1177/0002764214550299

Alexander DE (2008) Planning and management of complex emergencies. Disaster Plan Emerg Manag 75(11):20. https://doi.org/10.58729/1941-6687.1308

Ansell C, Gash A (2008) Collaborative governance in theory and practice. J Public Adm Res Theory 18(4):543–571. https://doi.org/10.1093/jopart/mum032

Armitage D, Marschke M, Plummer R (2011) Adaptive co-management and the paradox of learning. Glob Environ Chang 21(3):995–1001. https://doi.org/10.1016/j.gloenvcha.2011.04.002

Ballard HL, Belsky JM (2010) Participatory action research and environmental learning: implications for resilient forests and communities. Environ Educ Res 16(5–6):611–627. https://doi.org/10.1080/13504622.2010.505440

Benson D, Fritsch O, Langstaff L (2016) Local flood risk management strategies in England: patterns of application. J Flood Risk Manag 11:S827–S837

Berkes F (2009) Evolution of co-management: role of knowledge generation, bridging organizations and social learning. J Environ Manag 90(5):1692–1702. https://doi.org/10.1016/j.jenvman.2008.12.001

Berkes F (2010) Devolution of environment and resources governance: trends and future. Environ Conserv 37(4):489–500. https://doi.org/10.1017/S037689291000072X

Bonney R, Cooper CB, Dickinson J, Kelling S, Phillips T, Rosenberg KV, Shirk J (2009) Citizen science: a developing tool for expanding science knowledge and scientific literacy. Bioscience 59(11):977–984. https://doi.org/10.1525/bio.2009.59.11.9

Buytaert W, Zulkafli Z, Grainger S, Acosta L, Alemie TC, Bastiaensen J et al (2014) Citizen science in hydrology and water resources: opportunities for knowledge generation, ecosystem service management, and sustainable development. Front Earth Sci 2:26. https://doi.org/10.3389/feart.2014.00026

Chapagain D, Hochrainer-Stigler S, Velev S, Keating A, Hyun JH, Rubenstein N, Mechler R (2024) A taxonomy-based understanding of community flood resilience. Ecol Soc 29(4). https://doi.org/10.5751/es-15654-290436

Cornwall A (2008) Unpacking 'participation': models, meanings and practices. Community Dev J 43(3):269–283. https://doi.org/10.1093/cdj/bsn010

Danielsen F, Burgess ND, Jensen PM, Pirhofer-Walzl K (2014) Environmental monitoring: the scale and speed of implementation varies according to the degree of peoples' involvement. J Appl Ecol 51(5):1162–1168. https://doi.org/10.1111/1365-2664.12269

Devassy V (2025) Community-driven river resource management: perspectives from the Meenachil River Basin. Int J Multidiscip Res 7(1):1–5. https://www.ijfmr.com/papers/2025/1/37777.pdf

Dominelli L (2018) Green social work: from environmental crises to environmental justice, 2nd edn. Polity Press

Ensor J, Harvey B (2015) Social learning and climate change adaptation: evidence for international development practice. Wiley Interdiscip Rev Clim Chang 6(5):509–522. https://doi.org/10.1002/wcc.348

Fernandez-Gimenez ME, Ballard HL, Sturtevant VE (2008) Adaptive management and social learning in collaborative and community-based monitoring: a study of five community-based forestry organizations in the western USA. Ecol Soc 13(2):4. https://www.ecologyandsociety.org/vol13/iss2/art4/

Fiveable (2024) Flood management – AP environmental science. https://library.fiveable.me/key-terms/ap-enviro/flood-management

Fraisl D, Campbell J, See L, Wehn U, Wardlaw J, Gold M et al (2020) Mapping citizen science contributions to the UN sustainable development goals. Sustain Sci 15(6):1735–1751. https://doi.org/10.1007/s11625-020-00833-7

Gaillard JC, Mercer J (2013) From knowledge to action: bridging gaps in disaster risk reduction. Prog Hum Geogr 37(1):93–114. https://doi.org/10.1177/0309132512446717

Houston JB, Hawthorne J, Perreault MF, Park EH, Goldstein Hode M, Halliwell MR et al (2015) Social media and disasters: a functional framework for social media use in disaster planning, response, and research. Disasters 39(1):1–22. https://doi.org/10.1111/disa.12092

IIASA – International Institute for Applied Systems Analysis (n.d.) Revealing the power of citizen science for SDG advancement. https://iiasa.ac.at/news/jun-2023/revealing-power-of-citizen-science-for-sdg-advancement#:~:text=The%20achievement%20of%20the%20SDGs,complement%20and%20enhance%20official%20statistics

Intergovernmental Panel on Climate Change (2022) Climate change 2022: impacts, adaptation and vulnerability. In: Pörtner H-O, Roberts DC, Tignor M, Poloczanska ES, Mintenbeck K, Alegría A et al (eds) Contribution of working group II to the Sixth Assessment Report of the Intergovernmental Panel on Climate Change. Cambridge University Press. https://www.ipcc.ch/report/ar6/wg2/

Irwin A (1995) Citizen science: a study of people, expertise and sustainable development. Routledge. https://doi.org/10.4324/9780203202395

Josè Moisès D, Nyam YS (2025) Enhancing flood risk communication in Namibia: participatory flood early warning for early action. Environ Dev 57:10318. https://doi.org/10.1016/j.envdev.2025.101318

Joseph JK, Anand D, Prajeesh P, Zacharias A, George Varghese A, Pradeepkumar AP, Baiju KR (2020) Community resilience mechanism in an unexpected extreme weather event: an analysis of the Kerala floods of 2018, India. Int J Disaster Risk Reduct 49:101741. https://doi.org/10.1016/j.ijdrr.2020.101741

Kerala State Action Plan on Climate Change (2023–2030) (2022) Directorate of environment and climate change. Department of Environment, Government of Kerala

Kumar R (2018) Flood management in India. Open Access Int J Sci Eng 3(1):53–61

Lai L (2023) Exploring the main causes and effects of floods around the world. https://earth.org/what-are-the-main-causes-and-effects-of-floods/

Macherera M, Chimbari MJ (2016) A review of studies on community based early warning systems. Jàmbá J Disaster Risk Stud 8(1):Article 206. https://doi.org/10.4102/jamba.v8i1.206

Meier P (2015) Digital humanitarians: how big data is changing the face of humanitarian response. CRC Press. https://doi.org/10.1201/b18162

Mishra A (2021) Citizen engagement in disaster response: the case of Kerala floods. Asian J Soc Work 31(2):143–158

National Disaster Management Authority (NDMA) (2019) Guidelines on community-based disaster risk reduction. Government of India

Oluseyi OF, Akinbola G, Oloukoi J, Thonteh F, Enaruvbe G, Adagbasa E, Okarfor B (2011) Integrative approach of indigenous knowledge and scientific methods for flood risk analyses, responses and adaptation in rural coastal communities in Nigeria. Final project report for 2011 START grants for Global Change Research in Africa. http://start.org/download/gec11/fabiyi-final-report.pdf

Ostrom E (1996) Crossing the great divide: coproduction, synergy, and development. World Dev 24(6):1073–1087. https://doi.org/10.1016/0305-750X(96)00023-X

Pahl-Wostl C (2009) A conceptual framework for analysing adaptive capacity and multi-level learning processes in resource governance regimes. Glob Environ Chang 19(3):354–365. https://doi.org/10.1016/j.gloenvcha.2009.06.001

Palen L, Anderson KM (2016) Crisis informatics—new data for extraordinary times. Science 353(6296):224–225. https://doi.org/10.1126/science.aag2579

Pretty J (2003) Social capital and the collective management of resources. Science 302(5652):1912–1914. https://doi.org/10.1126/science.1090847

Rao D, Franklin ML (2024) Challenges in management of floods. 8th India Water week. https://indiawaterweek.in/IWW8_PaperFiles/WC-16/IWW8-SPD07_2-PAP-3-7307.pdf

Reed MS, Evely AC, Cundill G, Fazey I, Glass J, Laing A, Newig J, Parrish B, Prell C, Raymond C, Stringer LC (2010) What is social learning? Ecol Soc 15(4):1–10. https://www.ecologyandsociety.org/vol15/iss4/resp1/

Reuter C, Kaufhold MA (2018) Fifteen years of social media in emergencies: a retrospective review and future directions for crisis informatics. J Conting Crisis Manag 26(1):41–57. https://doi.org/10.1111/1468-5973.12196

Shaw R (2012) Community-based climate change adaptation in Vietnam: inter-linkages of environment, disaster, and human security. In: Shaw R, Pulhin J, Pereira J (eds) Community-based adaptation to climate change: scaling it up. Routledge, pp 107–138

Sukumaran ST, Birkinshaw SJ (2024) Investigating the impact of recent and future urbanization on flooding in an Indian river catchment. Sustainability 16(13):5652. https://doi.org/10.3390/su16135652

Sustainable Directory (2025) Community-led flood mitigation strategies in cities. Scenario. https://prism.sustainability-directory.com/scenario/community-led-flood-mitigation-strategies-in-cities/

Thomas J, George B, Suresh B (2020) Flood hazard assessment in Meenachil River Basin, Kerala. J Water Climate Change 11(4):1178–1192. https://doi.org/10.2166/wcc.2019.249

UNDRR (2015) Sendai framework for disaster risk reduction 2015–2030. United Nations Office for Disaster Risk Reduction

UNFCCC (2019) People-centred early warning systems: practices from climate adaptation projects. United Nations Framework Convention on Climate Change
Vincy MV, Brilliant R, Paul J, Pradeepkumar AP (2015) Comparison of riparian species diversity between the mainriver channel and subwatersheds of Meenachil river basin, Kerala, Southern India. Braz J Bot, 38(1):81–98
Wang L, Cui S, Li Y, Huang H, Manandhar B, Nitivattananon V, Fang X, Huang W (2022) A review of the flood management: from flood control to flood resilience. Heliyon 8(11):e11763. https://doi.org/10.1016/j.heliyon.2022.e11763
Wolff E, Schmitz S, Lazaro M (2021) People-centred approaches to flood early warning systems: a global review. Int J Disaster Risk Reduct 62:102398. https://doi.org/10.1016/j.ijdrr.2021.102398
World Meteorological Organization (2017) Integrated flood management tools series community based flood management. World Meteorological Organization

Chapter 7
Inclusion of Traditional Forest Resource Management Practices into Modern Micro-level Institutions: Decision-Making from an Eco-Social Work Perspective

Andria John and Laxmi Putran

Abstract This chapter emphasises leadership and decision-making while highlighting the integration of traditional and indigenous forest management techniques and their applicability in the modern micro-institution. This chapter advances eco-social work theory by demonstrating how tribal chief-led governance embodies ecological justice and can inform hybrid institutional models. It looks at how Eco-social work, which honours and empowers indigenous knowledge and practice, might facilitate that integration by promoting inclusive decision-making processes and participatory leadership and also outlines a way ahead towards more equitable and sustainable resource management, which will eventually benefit local communities. By focusing on leadership and decision-making processes that include traditional and indigenous knowledge systems, the chapter asserts the book's thrust on advancing social and environmental justice through inclusive and participatory approaches.

Keywords Eco-social work · Indigenous community · Decision making model

7.1 Introduction

Forest-dependent tribal communities in India, like the Kattunayakan, have historically maintained strong ecological, cultural, and spiritual relationships with forests. Known as a *Particularly Vulnerable Tribal Group (PVTG)*, the Kattunayakan community reside within the Wayanad Wildlife Sanctuary of Kerala. Forests serve purposes far beyond subsistence; they are deeply tied to identity, spirituality, and systems of governance. Most Kattunayakan households were nuclear, typically comprising two to four members. This demographic pattern places tribal chief leadership within tightly knit kinship networks.

A. John (✉) · L. Putran
Department of Social Work, Central University of Kerala, Kasaragod, Kerala, India

J. K. Joseph, A. P. Pradeepkumar (eds.), *Eco-Social Work for Disaster Risk Reduction*, Disaster Resilience and Green Growth, https://doi.org/10.1007/978-981-95-4885-9_7

Leadership within the Kattunayakan community is traditionally mediated through the t*ribal chief (Mudali/Mooppan in their dialect)* who has a crucial role in forest governance, decision-making, conflict resolution, and community representation. Tribal chief leadership is embedded in kinship, customary law, and traditional ecological knowledge. His authority is not only in forest resource allocation and protection but also in day-to-day life. Modern frameworks have attempted to decentralise forest management but often exclude or undermine the role of traditional institutions like the tribal chief. This creates a governance gap where community led practices are weakened by policy-driven exclusion. This chapter deals with the understanding of leadership and decision-making of tribal chief in forest management, integration of traditional governance into modern micro-level institutions, and the role of eco-social work in bridging these systems by promoting participatory and inclusive leadership models (Fig. 7.1). Within the disaster risk reduction (DRR) dialogue, Kattunayakan leadership (traditional model), could strengthen resilience in the situation of natural disasters.

Quantitative data includes the cross tabulations of tribal chiefs role in Non-Timber Forest Product (NTFP) management, dispute resolution, religious practices, and interactions with government stakeholders. Qualitative data include semi-structured interviews and focus group discussions capturing community perceptions of tribal chief leadership and traditional practices. The whole chapter is composed of qualitative and quantitative aspects to capture traditional vs. modern leadership dynamics.

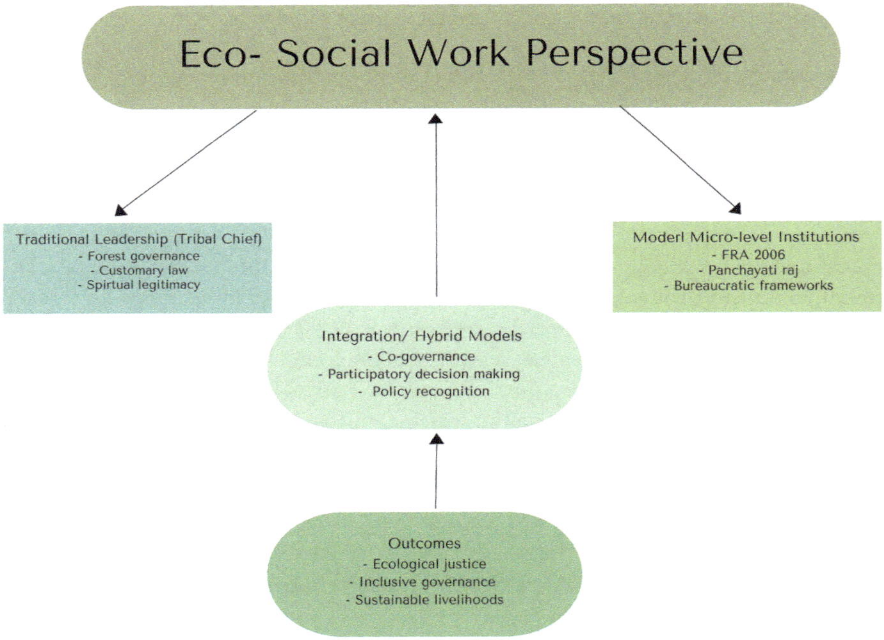

Fig. 7.1 Conceptual framework: integration of traditional and modern forest resource management through eco-social work perspective

7.2 Traditional Leadership and Decision-Making in Forest Resource Management

Traditional leadership among the Kattunayakan, a Particularly Vulnerable Tribal Group (PVTG), is centred on the tribal chief, the elder leader of the hamlet or community. Historically, the tribal chief is not only a symbolic head but also an active custodian of customary laws that regulate forest use, access, and conservation. As the primary authority in forest governance, the tribal chief embodies both political and spiritual legitimacy, derived from oral traditions and collective consensus (Ramachandran 1997; Government of Kerala 2013). The tribal chief's role in governance extends beyond mere authority: he mediates between individual households and the collective, ensuring equitable access to forest resources while discouraging exploitative extraction. This embedded leadership structure contributes to the sustainable management of forests, a practice deeply aligned with the PVTG worldview that regards forests as both livelihood and sacred space (Oldekop et al. 2020).

Figure 7.2 illustrates the various roles performed by the tribal chieftain (Mooppan), categorized according to their respective domains of strength. The chieftain has both ecological wisdom and cultural legitimacy. His leadership is shaped by inherited knowledge, taboos, and sacred norms that regulate sustainable usage of forest and its resources. Traditional forest governance practices include, Segment taboos (avoiding overharvesting in specific forest patches), Temporal taboos (restricting collection to certain seasons), and life-history taboos (prohibiting extraction of young plants or immature honeycombs). Such practices align with principles of ecological sustainability while reinforcing social cohesion. Chieftain's

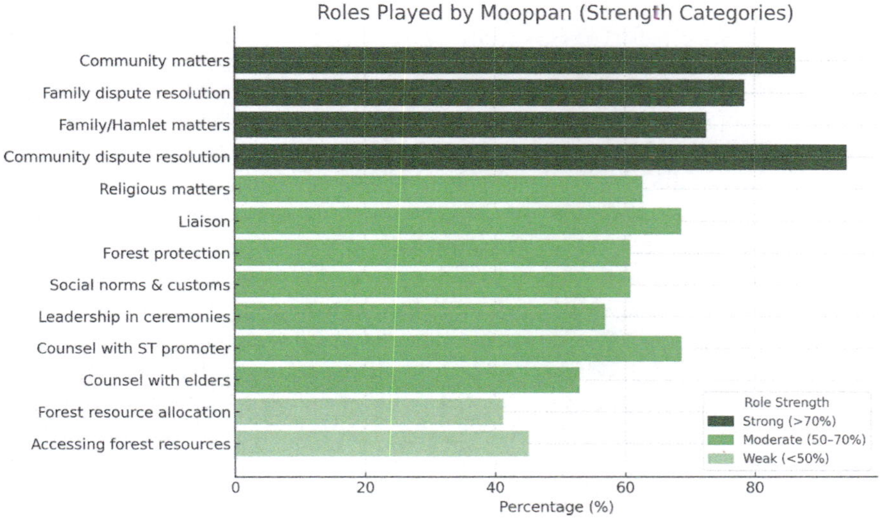

Fig. 7.2 Roles played by the tribal chieftain (Mooppan)—strength categories

authority is legitimised by collective compliance, in interviews, elders repeatedly pointed that "the chieftains word is binding" when it concerns forest use and access.

Thus, this chapter makes three distinct contributions to the literature. Primarily, it documents the leadership and decision-making processes of the Kattunayakan tribal chiefs using both qualitative and quantitative data, offering one of the few empirical accounts of a PVTG community in Kerala. Secondly it moves beyond descriptive ethnography by situating these traditional governance practices within the framework of eco-social work, thereby bridging indigenous knowledge with a global social work discourse that has rarely been applied to forest-dependent communities in India. Finally, this chapter proposes hybrid institutional models where tribal chief authority is not symbolic but operational in micro-level forest governance, advancing theoretical debates on participatory governance and ecological justice.

Altogether these contributions highlight how eco-social work can translate indigenous leadership into inclusive and sustainable forest management, addressing a gap in both policy design and academic research.

7.2.1 Tribal Chief's Role in Decision Making on Forest Resource Management

Dawson et al. (2021) found that locally controlled conservation efforts, particularly those where indigenous peoples and local communities have substantial decision-making influence, produce more positive outcomes for both biodiversity and well-being compared to externally controlled interventions. Traditional governance systems that incorporate customary institutions tend to be more effective than external approaches that supersede local practices. Leadership plays a crucial role in community forest governance, with Nath (2018) demonstrating that leadership significantly affects trade-offs between social and ecological outcomes through multiple causal pathways. However, the type of accountability matters: Kahsay et al. (2023) found that while bottom-up community monitoring showed no effect on forest outcomes, top-down governmental oversight improved conservation by enhancing leader accountability and crowding in community effort. Macura et al. (2013) emphasize the complexity of multilevel governance systems and the need for systematic evaluation of different governance modes' effectiveness.

The tribal chief's high involvement in community dispute resolution and family mediation underscores their role as custodians of social harmony and conflict resolution within indigenous systems (Table 7.1). This resonates with eco-social work's focus on strengthening community-led governance and justice in resource-dependent societies (Dominelli 2013; Gray et al. 2016). Their limited role in forest allocation and access highlights structural exclusions that eco-social work can address by advocating for equitable integration of customary practices into formal environmental governance (Berkes 2018).

Table 7.1 Roles played by the tribal chief

Domain	Level of role
Community dispute resolution	High
Community matters	High
Family dispute resolution	High
Family/hamlet matters	Medium
Liaison	Medium
Counsel with ST promoter	Medium
Religious matters	Medium
Social norms and customs	Medium
Forest protection	Medium
Leadership in religious ceremonies	Medium
NTFP management	Medium
Counsel with elders	Medium
Forest resource allocation	Low
Forest resource access	Low

Table 7.2 Mooppan category and role in NTFP management

Mooppan category	Role in NTFP management
Traditional	Medium
Dual	Medium
Mediator	Low

The Table 7.2, shows that higher-order tribal chiefs dominate decision-making on NTFP collection and conflict resolution, while household elders play only supportive roles. From an eco-social work perspective, the tribal chief's authority in regulating NTFPs illustrates how traditional leadership embodies community-driven accountability mechanisms that safeguard ecological sustainability and social equity. Traditional and dual-role of tribal chiefs' moderate engagement in NTFP decision-making reflects the cultural embeddedness of forest livelihoods. The absence of mediator Mooppan indicates uneven authority patterns, which eco-social work views as a risk for participatory justice (Ife 2016). Integrating mediator roles could democratize forest management and enhance sustainability by leveraging multiple knowledge systems (Berkes and Folke 2002).

7.2.1.1 Role in Forest Resource Allocation

Tribal forest resource management strategies demonstrate complex relationships between environmental stewardship, socioeconomic outcomes, and cultural preservation across diverse regional contexts. In West Bengal's Ajodhya Hill region, tribal communities like Santali and Munda face forest degradation from infrastructure development and tourism, forcing shifts from traditional forest dependency to alternative economic sectors, resulting in food insecurity and loss of medicinal plants (Karmakar and Mukherjee 2025). Similarly, Pacific Northwest tribes encounter social-ecological traps stemming from colonization, including restricted access to ancestral lands, declining resource quality, and erosion of traditional ecological knowledge, which have reduced both ecosystem resilience and tribal well-being (Long and Lake 2018). Climate change compounds these challenges by affecting water resources, aquatic species, and agricultural practices essential for tribal livelihoods and sovereignty (Cozzetto et al. 2013). However, tribal forest management systems offer valuable models for sustainable stewardship, providing approaches that balance multiple objectives while maintaining long-term environmental integrity and supporting indigenous communities, economically and culturally (Dockry and Hoagland 2017).

The data indicate that decisions on land and forest allocation remain concentrated with the hamlet tribal chief, reinforcing his central role in communal governance. These allocation roles reflect eco-social work's emphasis on distributive justice, as the tribal chief mediates competing claims to resources in ways that prioritize community well-being over market extraction. Resource allocation concentrated under dual-role Mooppan highlights both continuity and adaptation, balancing customary authority with evolving institutional demands (Table 7.3). Eco-social work situates this as a site for distributive justice, ensuring equitable access to forest resources while respecting indigenous governance norms (Dominelli 2018; Zapf 2009). This reflects the principle of linking environmental and social justice for community well-being.

7.2.1.2 Role in Forest Protection and Access

Indigenous forest governance reveals complex dynamics between traditional leadership, external actors, and conservation outcomes across Latin America. In Peru's Amazon, Kechwa-Lamas communities face pressure to adopt conservation-focused governance models rather than securing indigenous rights, potentially transforming farmers into conservationists under external influence (Egerlid et al. 2016). Traditional ecological knowledge systems demonstrate effective forest management approaches, as seen among Mapuche communities in Chile and tribal groups in India, emphasizing ecosystem processes over product extraction (Herrmann and Torri 2009). However, formal tenure rights do not automatically translate to practical benefits, with communities across Brazil, Bolivia, Nicaragua, and Guatemala encountering challenges including conflicts with other claimants, inadequate state support, and governance institution problems (Larson et al. 2008). In Panama's Comarca Ngäbe-Buglé, informal forest ownership systems operate within collective territories, where individual families control forest access and slow deforestation through de facto stewardship, highlighting hidden tenure layers that influence

Table 7.3 Tribal chief's role in forest resource allocation

Mooppan category	Role in resource allocation
Dual	Medium
Traditional	Low
Mediator	Low

Table 7.4 Tribal chief's role in forest protection

Mooppan category	Role in forest protection
Traditional	High
Dual	High
Mediator	Low

Table 7.5 Mooppan & resource access

Mooppan category	Role in resource access
Dual	Medium
Traditional	Medium
Mediator	Low

conservation outcomes (Smith et al. 2017). Table 7.4 describes the role of the tribal chief in forest protection, emphasizing their authority and responsibilities in safeguarding forest resources.

Chi-square results confirm a significant relationship between tribal chief category and authority in forest access and protection. Eco-social work interprets this as a form of ecological citizenship, where community compliance with the tribal chief reinforces environmental stewardship through culturally embedded legitimacy. Forest protection entrusted mainly to traditional and dual-role Mooppan shows the anchoring of ecological stewardship in cultural authority. Eco-social work underscores such indigenous custodianship as central to advancing climate justice and environmental sustainability (Gray and Coates 2013; Berkes 2018). The exclusion of mediator Mooppan points to governance gaps where inclusive eco-social practices could expand community resilience.

The moderate role of traditional and dual-role Mooppan in resource access decisions illustrates embedded systems of regulation that balance sustainability and community needs (Table 7.5). Eco-social work frames this as participatory governance rooted in local context, aligning with global calls for bottom-up environmental justice (Dominelli 2012; Ife 2016). Strengthening mediator involvement could democratize these decisions further. From global literature, Table 7.6 summarizes the tribal chief's role in forest resource management decision-making as a critical leadership function that integrates traditional ecological knowledge with contemporary management practices.

7.3 Traditional Leadership in Religious and Community Governance

7.3.1 Role in Religious Matters

Kattunayakan community in Tamil Nadu demonstrates complex traditional practices and leadership structures. The name "Kattunayakan" derives from "Kadu" (forest) and "Nayakkan" (leader or headman), literally meaning "King of the Jungle" (Gopi and Subramaniyam 2022). These forest-dwelling communities

Table 7.6 Summary of tribal chief's role in forest resource management decision-making

Category	Literature	Observed role of tribal chief	ESW interpretation
Forest resource management	Dawson et al. (2021): locally controlled conservation yields better socio-ecological outcomes. Nath (2018): leadership influences trade-offs. Kahsay et al. (2023): accountability pathways matter. Macura et al. (2013): multilevel governance complexity.	Tribal chiefs hold central authority, regulate access, resolve disputes, and maintain collective compliance.	Chiefs embody ecological citizenship, ensuring community-driven accountability and legitimacy in governance.
NTFP management	Field data (Table 7.2): Higher-order chiefs dominate NTFP collection and conflict resolution. Elders play supportive roles.	Regulate sustainable harvesting, enforce seasonal/spatial taboos, and mediate disputes.	Authority represents community-led accountability safeguarding equity and ecological sustainability.
Forest resource allocation	Karmakar and Mukherjee (2025): livelihood shifts under external pressures. Long and Lake (2018); Cozzetto et al. (2013): colonization & climate impacts. Dockry and Hoagland (2017): tribal stewardship models.	Hamlet chiefs control land and resource allocation; prioritize collective needs over market-driven extraction.	Reflects distributive justice by balancing competing claims to ensure community well-being.
Forest protection and access rights	Egerlid et al. (2016); Herrmann and Torri (2009); Larson et al. (2008); Smith et al. (2017): Indigenous tenure systems and traditional knowledge sustain ecosystems.	Chiefs determine access rules, enforce forest protection, and mediate external pressures.	Interpreted as ecological stewardship and cultural legitimacy reinforcing compliance and conservation.

maintain strong connections to traditional spiritual and healing practices, with their healthcare-seeking behaviour deeply embedded in magico-religious beliefs and indigenous medicines (Gandhi et al. 2017). The Kattunayakan possesses extensive ethno-medicinal knowledge, utilizing approximately 60 medicinal plants and parts for treating various ailments through their traditional healing systems (Amuthavalluvan 2011). Despite being historically nomadic, they continue following traditional customs and rituals (Gopi and Subramaniyam 2022). Along with other PVTGs like Paniyas, Todas, Kotas, Kurumbas, and Irulas, the Kattunayakan face declining or static population numbers and limited access to modern healthcare facilities, making their traditional practices particularly significant for community health and cultural preservation (Ganesh et al. 2021).

Religious roles concentrated among traditional and dual-role Mooppan highlight the intertwining of spirituality, ecology, and social order. Eco-social work interprets this as cultural-spiritual leadership reinforcing ecological ethics (Zapf 2009; Gray et al. 2016). These ceremonies sustain ecological consciousness and resilience,

Table 7.7 Tribal chief's role in religious ceremonies

Mooppan category	Role in religious ceremonies
Traditional	High
Dual	Medium
Mediator	Low

linking cultural continuity with sustainable practices. Table 7.7 demonstrate that the tribal chief's authority extends into the spiritual domain, reinforcing his legitimacy as both leader and ritual specialist.

7.3.2 Role in Community Governance

Research on the role of tribal chiefs in community governance shows their significant influence in rural development and local administration. In South Africa, chiefs are central to initiating development projects, mobilizing stakeholders, and ensuring compliance with government regulations, with communities perceiving them as more attuned to local needs than elected councillors (Mazibuko and Thebe 2024). In Ghana, chiefs play a critical role in socio-economic and cultural development, and their involvement in local government is seen as essential for effective administration (Asamoah 2019). The integration of chiefs into modern resource management systems led to role conflicts with local government actors, suggesting a need for consensus building to enhance policy effectiveness (Honyenuga and Wutoh 2019). In South Africa, traditional authorities has absolute decision-making power, particularly regarding benefit distribution, underscoring their pivotal role in community-based development initiatives (Ivanovich 2015).

Similarly, eco-social work underscores the importance of culturally legitimate leadership for inclusive governance, suggesting that tribal chief authority like African chieftaincy can be integrated into statutory systems without erasing indigenous legitimacy.

7.3.2.1 Role in Community Matters

In post-apartheid South African Research on traditional leadership shows complex dynamic between chiefs and democratic governance structures. Traditional chiefs continue to be regarded as ultimate authorities for natural resource regulation, particularly in firewood harvesting, despite the establishment of democratic institutions. Overall resource management appears weakened, attributed not only to confusion over leadership roles but also to socio-political factors including political expediency (Findlay and Twine 2018). The democratisation process faces significant challenges as land administration powers remain with unaccountable chiefs and headmen rather than elected officials, raising questions about compatibility between democratic decentralization and inherited traditional authority (Ntsebeza

Table 7.8 Tribal chief's role in hamlet governance

Governance area	Tribal chief involvement
Land disputes	High
Marriage negotiations	High
Work distribution	Moderate
Household conflicts	High

et al. 2004). This institutional ambiguity has undermined chiefs' control over natural resources as democratic local governance prioritises infrastructural development over resource management. Despite ongoing power struggles between chiefs and elected councillors, potential exists for collaboration in areas of common interest to improve natural resource management (Mwalukomo and Patel 2012).

Almost all Mooppan with varied roles, engage in dispute resolution, underscoring collective responsibility in sustaining community cohesion (Table 7.8). Eco-social work situates this as a practice of social inclusion, participatory justice, and resilience building (Dominelli 2013; Ife 2016). This universality makes dispute resolution a crucial entry point for integrating eco-social interventions with indigenous governance.

7.3.2.2 Role in the Village Council

The role of tribal chiefs in village councils, particularly in South Africa, highlights their continued significance in rural governance and development. Table 7.9 summarizes the tribal chief's role in the village council as the principal leader responsible for resolving disputes and maintaining social order. Chiefs are often seen as central figures in initiating development projects, mobilizing stakeholders, and ensuring community compliance with government regulations, with communities perceiving them as more attuned to local needs than elected councillors (Mazibuko and Thebe 2024). In the context of rural tourism, traditional authorities wield decisive power over development decisions, especially regarding employment and benefit distribution, which can be contentious within communities (Ivanovich 2015). However, tensions exist between traditional leadership and municipal authorities, as seen in the Mgwalana Traditional Authority, where overlapping mandates hinder socio-economic development unless collaborative frameworks are followed (George and Binza 2011). Despite debates about their relevance, traditional leaders remain widely accepted in rural areas and are recognized in the South African Constitution as important custodians of indigenous traditions and justice (Bayat and Ntonzima 2012).

Traditional tribal chiefs play crucial roles in village councils across various Indian states, serving as custodians of indigenous governance systems while adapting to modern democratic frameworks. Among the tribes of Arunachal Pradesh, the traditional council system called Jokom handles social and judicial matters, though it faces challenges from modern institutions like Panchayati Raj and religious conversion (Taba 2024). The Kuki chieftaincy in Manipur, led by hereditary chiefs known as Haosa, traditionally oversees land management and dispute resolution but

Table 7.9 Tribal chief in village council

Function	Tribal chief participation
Represents hamlet	High
Resolves inter-hamlet disputes	High
Negotiates with government	Moderate

confronts pressures from modernization and legal reforms (Ph Dominic Seiminlal Mate 2025). In Meghalaya's West Khasi Hills, the Dorbar system faces governance challenges due to illiteracy and poverty, with tribal elites potentially deviating from customary laws (Nongrum and Dash 2021). The Karbi of Assam demonstrate successful integration, where traditional chieftainship coexists symbiotically with the modern Karbi Anglong Autonomous Council, with both institutions supporting each other while preserving cultural identity (Das and Gogoi 2025).

7.3.3 Role in Dispute Resolution and Social Norms

Traditional dispute resolution systems play crucial roles in maintaining social harmony across indigenous communities in South Africa and India. In South Africa, tribal societies rely on indigenous conflict resolution systems rooted in cultural values of peace, tolerance, and solidarity to address various conflicts including territorial disputes, inheritance issues, and matrimonial problems (Masenya 2021). These traditional methods prioritize reconciliation and maintaining social relationships over mere case settlement, forming integral parts of time-proven social systems (Choudree 1999). The systems are especially oriented toward reconciliation and improvement of social relationships, though they face challenges due to unclear policy frameworks (Masenya 2021). In India, village disputes are primarily handled by panchayats (teams of five male elders) or individual male elders, who employ more assertive approaches compared to Western mediators (Alok et al. 2023). These traditional leaders effectively manage community disputes while upholding sociocultural norms and values that remain integral to organized society.

The Kattunayakan community maintains traditional customs and rituals in their nomadic lifestyle (Gopi and Subramaniyam 2022). However, the specific role of tribal chief in dispute resolution is not documented in the available literature. Other indigenous communities demonstrate similar traditional mediation systems. In Kerinci, the Depati Niniak Mamak serves as a customary mediator, with their capacity and disputing parties' motivation being key success factors, while emotional tensions and limited customary law understanding pose challenges (Gunawan 2021). The Tuha Peut institution in Aceh resolves village-level disputes, though knowledge gaps limit their effectiveness. Similarly, Mamak Limo Suku in Minangkabau communities actively mediate marital disputes, providing quick resolutions while avoiding formal court procedures (Fauzi and Eva 2025). Eco-social work highlights these practices as restorative rather than punitive, aligning with its principles of reconciliation, solidarity, and community cohesion.

7.3.3.1 Role in Family and Community Disputes

Indigenous conflict resolution systems demonstrate efficacy in building peace and promoting social solidarity, addressing disputes ranging from territorial expansion and resource competition to matrimonial conflicts and theft (Masenya 2021). In Sierra Leone, chieftaincy-based community dispute resolution has proven instrumental in post-conflict reconstruction, with chiefs successfully mediating various community disputes based on historical norms (Miller 2013). Similarly, traditional authorities among the Agni-morofouè people of Côte d'Ivoire serve as credible actors in peace construction and social cohesion building (Adou 2024). However, effectiveness depends critically on stakeholder acceptance of resolution approaches. Research from Ghana's northern conflicts shows that when dispute resolution methods align with local norms and customs, they achieve greater success, while misalignment leads to rejection and continued conflict (Tseer et al. 2023).

In the Kattunayakan community, the tribal chiefs mediate family-level conflicts (inheritance, domestic disputes) as well as inter-family quarrels. Their judgments, though unwritten, are binding. Higher-level tribal chiefs are consistently more involved in dispute resolution (Table 7.10).

7.3.3.2 Role in Social Norms and Customs

Research on indigenous communities shows complex dynamics between traditional leadership and social change across various tribal contexts (Table 7.11). Traditional customs and norms serve as behavioural guidelines that maintain social harmony and strengthen community solidarity, though they face challenges from modernization and globalization (Turyani et al. 2024). Social changes in indigenous communities occur due to demographic shifts, policy changes, new innovations, improved transportation, and evolving customary institutions, creating both positive and negative impacts (Nofianti 2020). Traditional leaders play crucial roles as peace judges and customary decision-makers, with their decisions being binding on disputing parties and aimed at creating family harmony (Nandigama, 2020). However, generational gaps in understanding traditional customs present significant challenges, particularly among younger generations who are vital for preserving cultural identity. The research emphasizes that renewal and adaptation of customs are necessary for maintaining relevance in modern contexts, requiring dialogue and consensus-building between younger generations, elders, and traditional leaders. Tribal chiefs shape social norms, customs, and taboos around forest use, gender participation in rituals, and marriage rules. This authority reinforces both ecological and social cohesion.

Table 7.10 Role in community dispute resolution

Mooppan category	Role in community dispute resolution
Dual	High
Traditional	High
Mediator	High

Table 7.11 Summary of tribal chief's role in social aspects of the community

Category	Literature	Observed role of tribal chief	ESW interpretation
Religious practices & ceremonies	Kattunayakan spiritual identity rooted in forests (Gopi and Subramaniyam 2022). Use of ethnomedicine and ritual healing (Amuthavalluvan 2011; Gandhi et al. 2017).	Acts as religious decision-maker and priest; presides over ceremonies and ritual authority (Tables 7.6 and 7.7).	Authority extends into spiritual ecology, legitimizing leadership and reinforcing cultural continuity.
Community governance	African & Indian chieftaincy studies: chiefs mobilize development, regulate resources, and resolve disputes (Mazibuko and Thebe 2024; Asamoah 2019).	High involvement in land disputes, marriage negotiations, household conflicts; moderate in work distribution (Table 7.7).	Represents culturally legitimate governance, balancing justice, order, and resource equity.
Village councils & inter-hamlet affairs	Chiefs remain central to governance despite modern institutions; co-existence with statutory systems in India & Africa (Ntsebeza et al. 2004; Das and Gogoi 2025).	Represents hamlet, negotiates with government, and resolves inter-hamlet disputes (Table 7.9).	Functions as bridge leaders, linking indigenous legitimacy with formal governance structures.
Dispute resolution (family & community)	Customary mediation prioritizes reconciliation and cohesion (Masenya 2021; Miller 2013).	Mediates inheritance, marital, and household disputes; unwritten judgments hold binding legitimacy.	Embodies restorative justice, fostering solidarity and reconciliation over punitive approaches.
Social norms & customs	Traditional leaders sustain norms but face pressures of modernization (Turyani et al. 2024; Nofianti 2020).	Shapes taboos on forest use, gender roles, and marriage rules; maintains collective identity.	Reinforces ecological and social cohesion, intergenerational responsibility, and cultural sustainability.

Thus, the tribal chief makes majority of the decision making, regarding restrictions on forest resource usage, allocation of forest resources, and decision on access to forest resources, dispute resolution, and social norms.

7.4 Challenges and Barriers to Integrating Traditional and Modern Governance

Traditional leadership approaches to forest resource management fundamentally differ from modern institutional strategies in their sources of legitimacy, governance mechanisms, and conservation effectiveness. Traditional systems derive their authority from culturally embedded institutions such as clan-based councils, village chiefs, and secret societies that enforce resource access through customary rules and social or spiritual sanctions, resulting in high local acceptance and compliance as demonstrated in Tanzania's Sonjo communities where 97% of households understood local forest regulations (Strauch et al. 2016). In contrast, modern institutional strategies rely on formalized structures including written regulations, state-led governance, and committee-based management that often struggle with legitimacy when imposed without harmonizing with established customs (Brown and Lassoie 2010; Zulu 2008; Kimengsi et al. 2022). While traditional approaches typically feature clear, locally recognized resource access rights enforced through collective monitoring embedded in daily social relations (Poudel 2017; Ostrom and Nagendra 2007), modern strategies frequently create confusion through overlapping administrative boundaries and suffer from elite capture, corruption, and rule evasion when lacking internal community recognition (Brown and Lassoie 2010; Zulu 2008; Poudel 2017). The conservation outcomes reflect these differences: traditional management systems generally achieved better forest conditions with less illegal logging (Persha and Blomley 2009; Strauch et al. 2016; Kajembe et al. 2003), while modern committee-led approaches often resulted in unsustainable management and widespread degradation (Brown and Lassoie 2010; Zulu 2008), suggesting that the cultural compatibility and endogenous nature of traditional institutions provide more effective foundations for environmental conservation than externally imposed formal structures (Kimengsi et al. 2022; Becker and Ghimire 2003).

Eco-social work interprets these conflicts as structural exclusions that marginalize indigenous authority, reinforcing the need for hybrid models where tribal chief leadership is not symbolic but operational in forest governance.

7.4.1 Conflicts Between Traditional Leadership and Modern Institutions

Research on tribal communities in Kerala reveals significant conflicts between traditional and modern institutions affecting social policy implementation and community development. The Kattunayakan tribe faces severe sustainability challenges, with extremely low agricultural income and livestock income, compounded by wildlife conflicts that hinder traditional livelihood activities (Ramakrishnan et al. 2024). Implementation of the Forest Rights Act 2006 (FRA 2006) among the Kadars demonstrates ongoing ambiguity in property rights and displacement issues, creating distress and social tensions (Anitha et al. 2015). Social interfaces between formal institutions and traditional communities often lack mutual trust, with conflicts emerging when formal regulations contradict local cultural practices that sustain livelihoods (Sunil 2007). The root causes of these conflicts stem from both restrictive forest policies and poorly designed development programs that fail to integrate tribal communities' existential needs (Damodaran 2006). These institutional conflicts significantly impede effective policy implementation and sustainable community development. As Tables 7.7 and 7.8 show, tribal chief authority in rituals and hamlet governance can inform participatory models that eco-social work seeks to embed in statutory systems. Despite their centrality, tribal chiefs often face marginalisation in state-led forest governance frameworks. Government institutions seldom recognize tribal chief authority under formal policy regimes, leading to conflicts in joint forest management committees and disputes over FRA (2006) provisions (Land Conflict Watch 2023).

7.4.2 *Institutional Barriers to Inclusion*

Indigenous forest management reveals significant institutional barriers limiting traditional authorities' participation in forest management. Indigenous communities face exclusion from policy-making processes that threaten their traditional ecological knowledge and self-determination rights (Carson et al. 2018). In Cameroon, decentralization efforts have inappropriately replaced traditional authorities with community forest management committees lacking internal legitimacy and dominated by local elites, undermining historical accountability systems that included clan heads and village chiefs (Brown and Lassoie 2010). Similarly, Mozambique's Land and Forestry laws fail to recognize customary institutions as legitimate authorities, constraining effective community participation, particularly among women, while most rural communities remain unaware of relevant natural resource statutes (Watts 2008). These institutional constraints are compounded by limited monitoring capabilities in local knowledge systems and scientific forestry's lack of institutional flexibility (Klooster 2002). Success requires hybrid institutions that combine traditional authority with democratic accountability and culturally appropriate legitimacy requirements.

7.4.2.1 Policy Barriers

The Forest Rights Act (FRA) 2006 shows its impact on tribal governance mechanisms in India. While the Act aims to recognise forest rights and empower indigenous communities, its implementation reveals significant limitations in transforming existing power structures. In Wayanad, Kerala, the FRA achieved limited success due to narrowly construed participation, with most new participatory spaces remaining closed, confining decentralization to technocratic essentials rather than addressing underlying patterns of domination (Chemmencheri 2013). The Act functions as a tool for subject-making of subaltern citizens, reducing discourse to recognition politics without substantial redistribution (Chemmencheri 2015). In Himachal Pradesh, powerful stakeholders used forest projects to undermine progressive FRA provisions, exacerbating institutional issues around inclusiveness and transparency (Aggarwal 2020). However, in Jharkhand, the FRA has shown positive outcomes with increased forest cover, though implementation challenges persist due to conflicts with existing governance structures like the Panchayati Raj Act (Dayal and Sharma 2024).

The FRA 2006 and Panchayati Raj Institutions sometimes bypass traditional leaders, imposing external governance structures. This results in overlapping authority and reduced recognition of tribal chief legitimacy.

7.4.2.2 Bureaucratic and Legal Challenges

Traditional authority in Africa reveals significant bureaucratic and legal challenges facing chiefs in modern governance systems. In post-apartheid South Africa, democratic decentralization has created tensions between elected local government and traditional authorities, with chiefs retaining land administration powers despite lacking democratic accountability (Ntsebeza et al. 2004). However, African politicians sometimes strategically cede resource control to traditional leaders to mobilize electoral support from non-coethnic groups, particularly regarding land allocation (Baldwin 2014). The codification of customary law presents additional challenges, as seen in Niger where chiefs must translate flexible, implicit local tenure arrangements into rigid legal frameworks, determining which customary practices should gain legal primacy (Lund and Hessling 1999). In Kenya, chiefs represent state bureaucracy rather than traditional authority, continuously negotiating legitimacy within pluralistic local authority structures while serving as crucial links between state and citizens (Osborn 2020). These cases demonstrate how oral traditions and customary legitimacy struggle within formal statutory frameworks. Bureaucratic red tape often side-lines tribal chiefs in decision-making. Their oral traditions and customary legitimacy lack legal codification, limiting their ability to claim recognition within statutory frameworks.

7.4.3 Tribal Chief's Interaction with External Stakeholders

Historical analysis shows that tribal communities, originally forced into forests by Aryan invasions, have maintained independent lifestyles despite lacking access to technology, education, and capital, with inevitable interactions emerging due to industrialization and globalization. In India, 750 diverse tribal groups across 27 states and 6 union territories exist outside mainstream Hindu society due to their distinctive culture and geographical isolation (Soren 2022). Stakeholder interactions significantly impact development interventions, as demonstrated in Pakistan's forestry projects where poor interaction between forest departments and local communities resulted from departmental supremacy in decision-making (Wattoo et al. 2010). Anti-corporate mobilization reveals additional complexity, with national advocacy networks playing crucial roles in resistance movements, though conflicts arise between actors with disparate goals, as evidenced in India's Niyamgiri bauxite mining case (Kraemar et al. 2013).

7.4.3.1 Counsel with Scheduled Tribe Promoters and Government Officials

Power dynamics between tribal chiefs and government representatives reveals complex challenges in decision-making processes for Scheduled Tribe communities. Political representation through reserved seats in decentralized governance institutions like Panchayati Raj Institutions provides necessary but insufficient conditions for tribal empowerment, as supportive institutional arrangements remain lacking (Patnaik 2013). The politics of inclusion demonstrates that state governments often make tribal identification decisions based on political considerations rather than established criteria, while legitimizing state control over tribal resources through welfare projects (Ambagudia 2011). Studies of co-management arrangements show that while power asymmetries exist between government officials and community members, deliberate participatory approaches can overcome these challenges through multiple-level decision-making processes (Espada and Kainer 2024). However, skewed power relations persist in local governance, indicating that inclusion alone cannot address historical exclusion from political processes (Vijayalakshmi 2002). Effective empowerment requires addressing structural power imbalances beyond mere representation. Among the Kattunayakan community, the tribal chiefs maintain active counsel with government representatives and Scheduled Tribe Promoters. However, these interactions are often asymmetrical, with formal institutions holding greater authority.

7.4.3.2 Counsel with Elders

Ritual specialists and community leaders negotiate authority and legitimacy within various social structures. Miller (1985) demonstrates that alternative communities use ritual structures to legitimize themselves by symbolically reconciling contradictions between public claims and daily realities. Snyder (1997) shows how among the Iraqw of Tanzania, male elders' ritual authority in the masay ceremony has become increasingly dependent on legitimation by elder women, as changing political-economic contexts have limited elders' secular authority to primarily ritual domains. Colonial states and ruling classes used ritualized negotiations with tribal populations to legitimize power, while common people reworked these practices to challenge dominant authority. Sihlé (2018) illustrates how Tibetan Buddhist tantrists collectively assess and adapt their large-scale rituals through institutionalized "elders' meetings," demonstrating expertise in managing sociological dynamics to maintain their collective existence and ritual effectiveness. Internally, among the Kattunayakan community the tribal chiefs rely on counsel from hamlet elders, ritual specialists, and kinship networks, reinforcing communal legitimacy. Elder advice frequently shapes final decisions, balancing hierarchy with collective governance.

7.5 Eco-Social Work Approaches to Bridging Traditional and Modern Governance

7.5.1 The Role of Eco-Social Work in Forest Resource Management

Eco-social work theory fundamentally transforms traditional leadership and governance by challenging modernist assumptions and promoting holistic, ecocentric approaches. Park (2025) introduces eco-transformational leadership as a radical departure from conventional transformational leadership, adapting its four attributes to embrace ecological perspectives rather than anthropocentric business models. This framework addresses the conspicuous absence of leadership discussions in eco-social work literature and supports practitioners in adopting eco-social practices. The transformation extends beyond individual leadership to organizational and community contexts. Boetto (2018) argues for shifting from modernist foundations to holistic ones that emphasize interdependence with the natural world, requiring perspective transformation and critical reflexivity in practice standards and codes of ethics. Powers et al. (2019) demonstrate how social work educators lead institutional and community transformation through infusion and integration models of eco-social education. Lysack (2012) illustrates capacity-building approaches that transform passive consumers into active environmental citizens, particularly within faith communities, highlighting eco-social workers' role in facilitating leadership emergence for earth advocacy.

Eco-social work, as conceptualized by Dominelli (2012), positions environmental justice as central to social work practice, emphasizing the inseparability of ecological and social systems. Within the Kattunayakan context, this approach provides a lens to integrate tribal chief-led traditional governance with contemporary forest management structures. By foregrounding principles of participation, justice, and interdependence, eco-social work offers strategies for rebalancing the power asymmetries between state-led interventions and indigenous autonomy.

Next we examine the intersection of environmental protection and social justice through indigenous practices and ecological knowledge. Baviskar (1997) examines how environmental discourse in India, particularly Ecological Marxism, attempts to merge social justice with ecological sustainability among tribal communities, though revealing tensions between activist and tribal perspectives on nature-development relationships. Javed et al. (2024) demonstrate how social justice theory applies to tribal women affected by hydropower projects, showing that tribal affiliation moderates the relationship between social justice dimensions and social sustainability outcomes. Both Ikeke (2024) and Shein and Sukinarhimi (2021) highlight how indigenous taboos serve as mechanisms for environmental protection and ecosystemic justice. Ikeke (2024) argues that Urhobo environmental taboos promote eco-justice but face threats from modernization, while Shein and Sukinarhimi (2021) show how Rukai water-related taboos maintain human-nature balance through core values of contamination prevention, resource sharing, and conservation, demonstrating traditional ecological knowledge as foundational to sustainable resource management.

Participatory governance among forest-dependent indigenous communities demonstrates how local practices intersect with broader governance mechanisms. Mistry et al. (2016), for example, showed how the *within-system viability framework* enabled Guyana's North Rupununi communities to develop community-owned solutions, anchored in indigenous knowledge and partnerships. Yet, participatory mechanisms are not without challenges. Leifsen et al. (2018) documented how state- and corporate-led participatory processes often fail to empower marginalized groups, compelling communities to adopt strategies to regain control over decision-making. Gender exclusion remains an enduring concern, as Rout (2017) highlighted in participatory forestry in India, where women's involvement was frequently underrepresented. Similarly, Ananda et al. (2025) demonstrated that in the Dayak Deah communities, participatory practices drew strength from traditional values such as *gotong royong*, with cultural identity and historical governance experiences shaping participation.

Eco-social work frameworks recognize these dynamics and advocate for embedding indigenous knowledge systems, gender inclusion, and community autonomy into participatory governance, thereby ensuring ecological sustainability is pursued alongside social justice.

Eco-social work interventions can promote participatory mechanisms to address gender and intergenerational inequalities through several key approaches. Social network analysis emerges as an innovative participatory tool that can analyse distinctive roles of men and women, involve different social groups, and examine patterns of community participation while promoting gender and social equity (Mertens et al. 2006). Participatory action research with practitioners shows that eco-social work interventions can be implemented across personal, individual, group, and organizational levels, though structural constraints at broader levels remain challenging (Boetto et al. 2020). In vulnerable communities, eco-social models demonstrate potential for promoting social transformation by respecting community rhythms and promoting individual empowerment, particularly important given that social vulnerability amplifies exposure to environmental threats (Belchior Rocha 2018). These approaches collectively suggest that participatory mechanisms in eco-social work can effectively address inequalities by engaging diverse community members in collaborative research and intervention processes.

A compelling case for transformative eco-social work practice that addresses interconnected environmental and social challenges can be made out. Dominelli (2013) advocates for "green social work" rooted in environmental justice, emphasizing practitioners' role in supporting marginalized populations who bear disproportionate environmental burdens despite minimal ecological footprints. This approach recognizes the interdependence between human well-being and ecosystem health. Boetto (2017) proposes a transformative eco-social model that challenges social work's modernist assumptions, arguing for holistic practice that addresses the profession's paradoxical role in sustaining industrial capitalism while promoting social justice. Dominelli (2014) demonstrates how environmental crises exacerbate existing inequalities, using post-tsunami Sri Lanka as evidence for integrating environmental justice into contemporary practice. Rambaree et al. (2019) conceptualize eco-social work as a paradigmatic shift from anthropocentric to eco-social perspectives, emphasizing interconnectedness of all life and requiring critical examination of societal structures that perpetuate social and ecological injustices through overconsumption and exploitation.

Case Studies of Successful Community-Driven Conservation Models
Community-led forest conservation initiatives show varied but generally positive outcomes. A global analysis of 643 community forest management cases found improvements in both environmental and income outcomes, though formalization sometimes negatively impacted forest access and resource rights (Hajjar et al. 2021). Successful models often shared key factors: secure tenure rights, supportive cultural institutions, and strong capacity-building mechanisms (Brooks et al. 2012).

Indigenous sovereignty and self-determination have emerged as critical elements. Denham (2017), examining Mexican cloud forest communities under a decade-long Payments for Ecosystem Services program, noted that attention to autonomy and cultural governance ensured both conservation and livelihood enhancement. Similarly, Novick et al. (2023) found that locally led approaches in Indonesia reduced forest loss and illegal activities, with higher community participation correlating with lower levels of poaching and encroachment.

These cases highlight how eco-social work can support indigenous leadership in creating pathways for sustainable, community-driven conservation. In the Kattunayakan context, strengthening the tribal chief's leadership role within participatory structures aligns with these global lessons.

7.5.2 Participatory Leadership Models for Integrating Indigenous Knowledge

Examples from India and global indigenous communities underscore the effectiveness of community-based co-governance. Eco-social work emphasizes collaborative leadership models that integrate local ecological knowledge with broader governance frameworks. For the Kattunayakan, this means positioning the tribal chief not as a marginal figure but as a co-equal actor in micro-level forest governance, ensuring decisions respect cultural taboos, belief systems, and ecological rhythms.

Eco-social work interventions prioritize capacity-building to empower indigenous leadership. Training initiatives could strengthen tribal chief-led institutions by enhancing negotiation skills with government actors, promoting gender-inclusive decision-making within village council, and building awareness of policy frameworks (e.g., FRA 2006) while maintaining cultural legitimacy. Such interventions align with Dominelli's vision of empowering local communities to shape environmental futures through participatory, justice-oriented governance.

7.5.3 Policy Recommendations for Inclusive Governance

Eco-social work perspectives advocate for reforms that formally acknowledge traditional leaders within statutory governance systems. This includes amending forest policies to explicitly recognize tribal chief authority in decision-making processes, thereby reducing institutional marginalization. Strengthening tribal chief leadership requires institutional mechanisms for joint decision-making between traditional leaders and formal governance bodies. Eco-social work's emphasis on participatory democracy supports this integration, ensuring that indigenous voices are not sidelined but actively shape conservation and development policies.

7.6 Conclusion

This study shows how tribal chiefs plays a central role in leadership and decision-making across forest resource management, religious matters, community governance, and dispute resolution. This chapter contributes uniquely to the literature in three ways. It provides rare empirical insights into the leadership and decision-making processes of a Particularly Vulnerable Tribal Group in Kerala; it frames these traditional governance practices through an eco-social work lens, a perspective seldom applied in the Indian tribal context; and it advances the idea of hybrid institutional models that operationalize rather than marginalize indigenous authority. These contributions extend current debates on participatory governance and ecological justice. However, challenges arise in integrating these traditional governance structures with modern forest institutions, largely due to policy and bureaucratic barriers. Eco-social work provides a transformative framework for bridging these divides, by aligning social justice with ecological justice and recognizing the cultural foundations of indigenous leadership. It offers a pathway to integrate tribal chief authority into participatory forest governance models that are inclusive, just, and sustainable. Future research should extend this analysis to other PVTG communities to explore the diversity of governance models and their interactions with modern institutions. Longitudinal studies are also needed to examine the long-term impacts of integrating traditional and modern governance, particularly in terms of ecological outcomes, community autonomy, and gender inclusion. The empirical data confirm that hamlet tribal chiefs retain consistently high authority across domains, NTFP management (Table 7.1), forest allocation (Table 7.3), and conflict resolution (Table 7.9). This pattern demonstrates that despite policy marginalization, tribal chief leadership remains the de facto governance structure for forest-dependent communities. Recognizing this authority through eco-social work frameworks allows for policy reforms that embed participatory, justice-oriented, and ecologically rooted governance in micro-level institutions.

The integration of traditional forest resource management practices into modern micro-level institutions provides a compelling case for transformative eco-social work practice that addresses the interconnected environmental and social challenges of our time. As Dominelli (2013) highlights, green social work is rooted in environmental justice, positioning practitioners to advocate for marginalized populations who disproportionately experience environmental burdens despite their minimal ecological footprints. This resonates strongly with indigenous and forest-dependent communities, whose knowledge systems and sustainable practices remain vital for ecological balance but are often overlooked in dominant governance structures. Boetto's (2017) transformative eco-social model further underscores the need for social work to critically engage with its paradoxical role in sustaining industrial capitalism while simultaneously pursuing social justice. This tension is directly reflected in the challenges faced by traditional forest governance systems, where community-based leadership and decision-making structures must negotiate with state-led, market-driven models of resource use. Integrating these systems through an eco-social lens fosters holistic practices that value indigenous ecological knowledge, reciprocity, and collective stewardship. Dominelli (2014), shows how environmental crises exacerbate existing inequalities, reinforcing the urgency of embedding environmental justice into social work. Similarly, tribal forest communities often face compounded vulnerabilities, ecological degradation, policy marginalization, and socio-economic exclusion, which demand responses that are both socially just and ecologically sustainable. Building on Jack et al. (2008), eco-social work represents not merely an expansion of practice but a paradigmatic shift from anthropocentric to eco-social perspectives. This shift requires practitioners and institutions to recognize the interconnectedness of all life forms and to critically examine societal structures that perpetuate ecological exploitation and social exclusion. Leadership and decision-making within indigenous forest management exemplify this interconnectedness, offering models of governance that embed ecological reciprocity, intergenerational responsibility, and community well-being.

In conclusion, bridging traditional forest resource management with modern micro-level institutions through eco-social work perspectives contributes to both theoretical innovation and practical transformation. It redefines leadership and decision-making as processes rooted in ecological justice, cultural respect, and sustainability, thus positioning eco-social work as a vital framework for addressing the intertwined crises of social inequality and environmental degradation. Thus, the eco-social work framework could help in implementing the DRR strategies among the forest dependent tribal communities like Kattunayakan, who are living in ecologically fragile zone, and prone to frequent natural disasters. Thus this chapter has attempted to connect the traditional tribal/indigenous resource management system as a model to be adapted to DRR strategies among the vulnerable groups.

References

Adou AD (2024) Traditional authorities, a reliable alternative for sustainable social cohesion and peace in Africa: The case of the Agni-Morofouépeople of Bongouanou (Côte d'Ivoire). Groupe Interdisciplinaire de Recherche sur les Cultures et les Identités. https://doi.org/10.61585/pud-girci-v1n301

Aggarwal A (2020) Improving forest governance or messing it up? Analysing impact of forest carbon projects on existing governance mechanisms with evidence from India. Forest Policy Econ 111:102080. https://doi.org/10.1016/j.forpol.2019.102080

Alok VN (2023) Panchayats and dispute resolution: A report. Ministry of Panchayati Raj. Indian Inst Public Adm.

Ambagudia J (2011) Scheduled tribes and the politics of inclusion in India. Asian Soc Work Policy Rev 5(1):33–43. https://doi.org/10.1111/j.1753-1411.2010.00047.x

Amuthavalluvan V (2011) Ethno medicinal practices and traditional healing system of Kattunayakan in Tamilnadu: an anthropological study. Int Multidiscip Res J 1(7). https://updatepublishing.com/journal/index.php/imrj/article/view/1497

Ananda R, Suryadi B, Arisanty D, Nuryadin MR (2025) Social construction of participation among Dayak Deah Indigenous Communities in village fund management = Konstruksi sosial partisipasi masyarakat adat Dayak Deah dalam pengelolaan dana desa. Society 13(1):747–764. https://doi.org/10.33019/society.v13i1.770

Anitha V, Balakrishnan R, Krishnakumar J (2015) Community participation in natural resource conservation: forest rights act among the *Kadars* in Kerala, India. J Environ Res Develop 9(3):669–680

Asamoah K (2019) A qualitative study of chieftaincy and local government in Ghana. Global J Soc Anthropol 8(12):001–006

Baviskar A (1997) Tribal politics and discourses of environmentalism. Contributions to Indian Sociology, 31(2), 245–268. https://doi.org/10.1177/006996697031002002

Baldwin K (2014) When politicians cede control of resources: land, chiefs, and coalition-building in Africa. Comp Polit 46(3):253–271. https://doi.org/10.5129/001041514810943036

Becker CD, Ghimire K (2003) Synergy between traditional ecological knowledge and conservation science supports forest preservation in Ecuador. Ecol Soc 8(1). http://www.consecol.org/vol8/iss1/art1/

Belchior Rocha H (2018) Social work practices and the ecological sustainability of socially vulnerable communities. Sustainability 10(5):1312. https://doi.org/10.3390/su10051312

Berkes F (2018) Sacred ecology. Routledge

Berkes F, Folke C (2002) Linking social and ecological systems. Cambridge University Press, Cambridge

Boetto H (2017) A transformative eco-social model: Challenging modernist assumptions in social work. Br J Soc Work, 47(1):48–67. https://doi.org/10.1093/bjsw/bcw149

Boetto H (2018) Advancing transformative eco-social change: shifting from modernist to holistic foundations. Aust Soc Work 72(2):139–151. https://doi.org/10.1080/0312407X.2018.1484501

Boetto H, Bowles W, Närhi K, Powers M (2020) Raising awareness of transformative ecosocial work: participatory action research with Australian practitioners. Int J Soc Work 29(4):300–309

Brooks JS, Waylen KA, Mulder MB (2012) How national context, project design, and local community characteristics influence success in community-based conservation projects. Proc Natl Acad Sci USA 109(52):21265. https://doi.org/10.1073/pnas.1207141110

Brown HCP, Lassoie JP (2010) Institutional choice and local legitimacy in community-based forest management: lessons from Cameroon. Environ Conserv 37(3):261–269. http://www.jstor.org/stable/44519980

Carson SL, Kentatchime F, Nana ED, Njabo KY, Cole BL, Godwin HA (2018) Indigenous peoples' concerns about loss of forest knowledge: implications for forest management. Conserv Soc 16(4):431–440. http://www.jstor.org/stable/26500657

Chemmencheri SR (2013) Decentralisation, participation and boundaries of transformation: Forest Rights Act, Wayanad, India. Commonw J Local Gov 12:51–68. https://doi.org/10.5130/cjlg.v12i0.3264

Chemmencheri SR (2015) State, social policy and subaltern citizens in adivasi India. Citizenship Studies 19(3–4):436–449. https://doi.org/10.1080/13621025.2015.1006579

Choudree RB (1999) Traditions of conflict resolution in South Africa. Afr J Confl Resolut 1:9–27

Cozzetto K, Chief K, Dittmer K et al (2013) Climate change impacts on the water resources of American Indians and Alaska Natives in the U.S. Clim Chang 120:569–584. https://doi.org/10.1007/s10584-013-0852-y

Damodaran A (2006) Tribals, forests and resource conflicts in Kerala, India: the status quo of policy change. Oxf Dev Stud 34(3):357–371. https://doi.org/10.1080/13600810600921976

Das S, Gogoi C (2025) From tribal chieftainship to autonomous council: understanding the intersectionality of governance in West Karbi Anglong, Assam, India. Altern Int J Indig People. https://doi.org/10.1177/11771801251358554

Dawson NM, Coolsaet B, Sterling EJ, Loveridge R, Gross-Camp ND, Wongbusarakum S, Sangha KK, Scherl LM, Phuong Phan H, Zafra-Calvo N, Lavey WG, Byakagaba PCJ, Idrobo A, Chenet NJ, Bennett SM, Rosado-May FJ (2021) The role of indigenous peoples and local communities in effective and equitable conservation. Ecol Soc 26(3):19. https://doi.org/10.5751/ES-12625-260319

Dayal SS, Sharma D (2024) Jharkhand tribal gram Sabha: forest governance dynamics under FRA 2006. ShodhKosh J Visual Perform Arts 5(6). https://doi.org/10.29121/shodhkosh.v5.i6.2024.3568

Denham D (2017) Community forest owners evaluate a decade of payments for ecosystem services in the Mexican cloud forest: the importance of attention to indigenous sovereignty in conservation. Soc Nat Resour 30(9):1064–1079. https://doi.org/10.1080/08941920.2017.1295495

Dockry MJ, Hoagland SJ (2017) A special issue of the journal of forestry—tribal forest management: innovations for sustainable forest management. J For 115(5):339–340. https://doi.org/10.5849/JOF-2017-040

Dominelli L (2012) Green social work: from environmental crises to environmental justice. [Book Review]. Br J Soc Work 42(8):1636–1637. https://doi.org/10.1093/bjsw/bcs183

Dominelli L (2012, 2013, 2018) Green social work. Polity Press

Dominelli L (2014) Promoting environmental justice through green social work practice: a key challenge for practitioners and educators. Int Soc Work 57(4):338–345. https://doi.org/10.1177/0020872814524968

Dominelli L (2018) Green social work in theory and practice: a new environmental paradigm for the profession. In: The Routledge handbook of green social work. Routledge, Abingdon, p 12

Egerlid J, Marquardt K, Bartholdson O (2016) Forest conservation versus indigenous forest territory rights in the Peruvian Amazon: the case of the Kechwa-Lamas village Alto Huaja and the roles of external actors. Int J Agric Resour Gov Ecol 12(4):381–405

Espada ALV, Kainer KA (2024) Decision making processes and power dynamics in timber production co-management: a comparative analysis of seven Brazilian Amazonian community-based projects. Forest Policy Econ 159:103121. https://doi.org/10.1016/j.forpol.2023.103121

Fauzi F, Eva Y (2025) The existence of customary role in the settlement of marriage disputes (a study of the role of Mamak Limo Suku in Jorong Sungai Sanda). Jurnal Kajian dan Pengembangan Umat 8(1):10.31869/jkpu.v8i1.6542

Findlay SJ, Twine WC (2018) Chiefs in a democracy: a case study of the 'new' systems of regulating firewood harvesting in post-apartheid South Africa. Land 7(1):35. https://doi.org/10.3390/land7010035

FRA (2006) is the short form of Scheduled Tribes and Other Traditional Forest Dwellers (Recognition of Forest Rights) Act, 2006 and interchangeably used asForest Rights Act 2006

Gandhi S, Verma VR, Dash U (2017) Health seeking behaviour among particularly vulnerable tribal groups: a case study of Nilgiris. J Public Health Epidemiol 9(4):74–83. https://doi.org/10.5897/JPHE2017.0911

Ganesh B, Thangarasu R, Acharya SK, Sridharan V (2021) Particularly vulnerable tribal groups of Tamil Nadu, India: A sociocultural anthropologicalreview. Indian J Public Health, 65(4):403–409. https://doi.org/10.4103/ijph.IJPH_2_21

George K, Binza MS (2011) The role of traditional leadership in promoting governance and development in rural South Africa: a case study of the Mgwalana traditional authority. J Public Admin 46:947

Gopi M, Subramaniyam K (2022) Social customs of the Kattunayakkan tribes in Nilakottai Taluk. Int J Res Human Soc Sci 9(1). https://doi.org/10.15613/hijrh/2022/v9i1/213624

Government of Kerala (2013) Report on the Socioeconomic Status.https://www.stdd.kerala.gov.in/sites/default/files/inline-files/surveyd_2008.pdf

Gray M, Coates J, Yellow Bird M, Hetherington T (Eds.) (2013) Decolonizing Social Work (1st ed.). Routledge. https://doi.org/10.4324/9781315576206

Gray M, Coates J, Hetherington T (2016) Environmental social work. Routledge, Abingdon

Gunawan A (2021) Conflict mediation based on local wisdom: study on customary dispute resolution in Pelompek Village, Gunung Tujuh, Kerinci. Muharrik:Jurnal Dakwah dan Sosial 4(2):237–254. https://doi.org/10.37680/muharrik.v4i02.996

Hajjar R, Oldekop JA, Cronkleton P, Newton P, Russell AJM, Zhou W (2021) A global analysis of the social and environmental outcomes of community forests. Nature Sustainability 4:216–224. https://doi.org/10.1038/s41893-2020-00633-y

Herrmann TM, Torri MC (2009) Changing forest conservation and management paradigms: traditional ecological knowledge systems and sustainable forestry: perspectives from Chile and India. Int J Sustain Dev World Ecol 16(6):392–403. https://doi.org/10.1080/13504500903346404

Honyenuga BQ, Wutoh EH (2019) Ghana's decentralized governance system: the role of chiefs. Int J Public Leader 15(1):2–18. https://doi.org/10.1108/IJPL-01-2018-0005

Ife J (2016) Community development in an uncertain world. Cambridge University Press, Cambridge

Ikeke MO (2024) Urhobo indigenous environmental taboos and the imperative of fostering ecosystemic justice. Afr J Clim Chang And, 3(1):272–283. https://doi.org/10.37284/ajccrs.3.1.2166

Ivanovich M (2015) The role of tribal authorities in rural tourism development in South Africa : the case of the kingdom of the rain queen. Afr J Phys Health Educ Recreat Dance 21:37–52

Jack BK, Kousky C, Sims KR (2008) Designing payments for ecosystem services: lessons from previous experience with incentive-based mechanisms. Proc Natl Acad Sci USA 105(28):9465–9470. https://doi.org/10.1073/pnas.0705503104

Javed T, Zhao X, Mahmoud AB, Fuxman L, Houjeir R (2024) Promoting social justice and women's rights: Mitigating the negative impacts of largehydropower projects on tribal women in the Hindu Kush Himalaya region. Environ Dev, 50:101001. https://doi.org/10.1016/j.envdev.2024.101001

Kahsay GA, Bulte E, Alpizar F, Hansen LG, Medhin H (2023) Leadership accountability in community-based forest management: experimental evidence in support of governmental oversight. Ecol Soc 28(4). https://doi.org/10.5751/ES-14469-280420

Kajembe GC, Luoga EJ, Kijazi MS, Mwaipopo CS (2003) The role of traditional institutions in the conservation of forest resources in east Usambara, Tanzania. Int J Sustain Dev World Ecol 10(2):101–107. https://doi.org/10.1080/13504500309469789

Karmakar S, Mukherjee S (2025) Development and sustainability: a case study on forest resource dependency of tribal communities in the Hilly Areas of Ajodhya Hill Region, Purulia, West Bengal. Int J Multidiscip Res 7(4):1–13. https://doi.org/10.36948/ijfmr.2025.v07i04.50523

Kimengsi JN, Owusu R, Djenontin IN, Pretzsch J, Giessen L, Buchenrieder G, Pouliot M, Acosta AN (2022) What do we (not) know on forest management institutions in sub-Saharan Africa? A regional comparative review. Land Use Policy 114:105931. https://doi.org/10.1016/j.landusepol.2021.105931

Klooster DJ (2002) Toward adaptive community forest management: integrating local forest knowledge with scientific forestry. Econ Geogr 78(1):43–70. https://doi.org/10.1111/j.1944-8287.2002.tb00175.x

Kraemar R, Whiteman G, Banerjee B (2013) Conflict and astroturfing in Niyamgiri: the importance of national advocacy networks in anti-corporate social movements. Organ Stud 34(5-6):823–852. https://doi.org/10.1177/0170840613479240

Land Conflict Watch (2023) Fortress walls: Rights and conflicts in protected areas (Quarterly data analytics report)

Larson AM, Cronkleton P, Barry D, Pacheco P (2008) Tenure rights and beyond: community access to forest resources in Latin America. Center for International Forestry Research (CIFOR), Bogor

Leifsen E, Gustafsson M-T, Guzmán-Gallegos MA, Schilling-Vacaflor A (2018) New mechanisms of participation in extractive governance: between technologies of governance and resistance work. In: Leifsen E, Gustafsson M-T, Guzmán-Gallegos MA, Schilling-Vacaflor A (eds) New mechanisms of participation in extractive governance. Routledge

Long JW, Lake FK (2018) Escaping social-ecological traps through tribal stewardship on national forest lands in the Pacific northwest, United States of America. Ecol Soc 23(2):10. https://doi.org/10.5751/ES-10041-230210

Lund C, Hessling G (1999) Traditional chiefs and modern land tenure law in Niger. In: African chieftaincy in a new socio-political landscape. Lit Verlag, Hamburg, pp 135–154

Lysack M (2012) Building capacity for environmental engagement and leadership: an ecosocial work perspective. Int J Soc Welf 21(3). https://doi.org/10.1111/j.1468-2397.2011.00854.x

Macura B, Secco L, Pullin AS (2013) Does the effectiveness of forest protected areas differ conditionally on their type of governance? Environ Evid 2(14). https://doi.org/10.1186/2047-2382-2-14

Masenya TM (2021) Indigenous conflict resolution systems: building cultural peace and social solidarity among indigenous communities in South Africa. In: Nicholson R (ed) Natural healing as conflict resolution. IGI Global Scientific Publishing, Hershey, pp 226–241. https://doi.org/10.4018/978-1-7998-3665-0.ch010

Mate D (2025) Kuki chieftainship in Manipur: Balancing tradition with modern governance. J Softw Eng Simul, 11(5):22–29. https://doi.org/10.35629/3795-11052229

Mazibuko O, Thebe V (2024) Identity, experience, and non-partisan: the institution of traditional chiefs and development in a rural local municipality in north West Province, South Africa. Afr Identities 23(3):683–695. https://doi.org/10.1080/14725843.2024.2324101

Mertens F, Saint-Charles J, Demeda K, de Castro M, Passos CJS, Lucotte M, Guimarães JRD, Mergler D (2006) Community network analysis for addressing gender, equity and participation in ecohealth research. In: Proceedings of IDRC'S participation in the 11th world congress on public health/8th Brazilian congress on collective health, Rio de Janeiro, pp 102–111

Miller G (1985) Work, ritual structures, and the legitimation of alternative communities. Work Occup. https://doi.org/10.1177/0730888485012001001

Miller WM (2013) Chieftaincy-based community dispute resolution: the case of Sierra Leone. Confl Resolut Q 30(3):295–308. https://doi.org/10.1002/crq.21066

Mistry J, Berardi A, Tschirhart C, Bignante E, Haynes L, Benjamin R, Albert G, Xavier R, Robertson B, Davis O, Jafferally D, de Ville G (2016) Community-owned solutions: identifying local best practices for social-ecological sustainability. Ecol Soc 21(2). https://www.jstor.org/stable/26270391

Mwalukomo H, Patel Z (2012) Chieftaincy and democratic local governance in rural South Africa: natural resources management in QwaQwa. Dev South Afr 29(2):259–272. https://doi.org/10.1080/0376835X.2012.675696

Nandigama S (2020) Performance of success and failure in grassroots conservation and development interventions: gender dynamics in participatory forest management in India. Land Use Policy 97:103445. https://doi.org/10.1016/j.landusepol.2018.05.061

Nath S (2018) Leadership and trade-offs between social–ecological outcomes in community-governed forests. Soc Nat Resour 31(7):790–806. https://doi.org/10.1080/08941920.2018.1448131

Nofianti T (2020) Perubahan Sosial Komunitas Suku Arfak Kabupaten Pegunungan Arfak Papua Barat. Agribus J. https://doi.org/10.15408/AJ.V13I2.13956

Nongrum R, Dash T (2021) Tribal's perception on governance of village council (Dorbar) system: A case study of West Khasi Hills District of MeghalayaState. BSSS J Soc Work, 13(1):34–68. https://doi.org/10.51767/jsw1304

Novick B, Crouch J, Ahmad A, Kartikawati SM, Sagita N, Miller AE (2023) Understanding the interactions between human Well-being and environmental outcomes through a community-led integrated landscape initiative in Indonesia. Environ Dev 45:100791. https://doi.org/10.1016/j.envdev.2022.100791

Ntsebeza L, Ribot J, Larson AM (2004) Democratic decentralisation and traditional authority: dilemmas of land Administration in Rural South Africa. Eur J Dev Res 16:71–89. https://doi.org/10.1080/0957881041000l688743

Ntonzima L, Bayat MS (2012) The role of traditional leaders in South Africa: A relic of the past, or a contemporary reality? Oman Chapter of ArabianJ Bus Manag Rev, 1(6). https://doi.org/10.12816/0002112

Oldekop JA, Rasmussen LV, Agrawal A, Bebbington AJ, Meyfroidt P,Bengston DN, Blackman A, Brooks S, Davies P, Dinsi SC, Fontana LB,Gumucio T, Kumar C, Kumar K, Moran D, Mwampamba TH, Nasi R,Nilsson MAM, Wilson SJ (2020) Forest-linked livelihoods in aglobalized world. Nat Plants, 6(12):1400-1407. https://doi.org/10.1038/s41477-020-00814-9

Osborn M (2020) Chiefs, elders, and traditional authority. In: Nic C, Karuti K, Gabrielle L (eds) The Oxford handbook of Kenyan politics. Oxford University Press, Oxford. https://doi.org/10.1093/oxfordhb/9780198815693.013.20

Ostrom E, Nagendra H (2007) Insights on linking forests, trees, and people from the air, on the ground, and in the laboratory. Proc Natl Acad Sci 103(51):19224–19231. https://doi.org/10.1073/pnas.0607962103

Park TK (2025) Eco-transformational leadership framework for social workers to promote climate and ecological justice. Br J Soc Work. https://doi.org/10.1093/bjsw/bcaf102

Patnaik P (2013) Does political representation ensure empowerment? Scheduled tribes in decentralised local governments of India. J South Asian Dev. https://doi.org/10.1177/0973174113476998

Persha, L. And Blomley, T. (2009), Management decentralization and montane Forest conditions in Tanzania. Conserv Biol, 23: 1485-1496. https://doi.org/10.1111/j.1523-1739.2009.01276.x

Poudel DP (2017) Are traditional and modern forestry institutions complementary or do they prescribe contradictory rights in trans-Himalayan Nepal? A lesson to be learned for REDD+ implementation. Nor Geogr Tidsskr 72(1):13–26. https://doi.org/10.1080/00291951.2017.1413418

Powers M, Schmitz C, Moritz MB (2019) Preparing social workers for eco social work practice and community building. J Community Pract 27(3-4):446–459. https://doi.org/10.1080/10705422.2019.1657217

Ramachandran B (1997) Ethnographic Notes on the Kattunaickens: A FoodGathering Tribe of Wayanad, Kerala. J Hum Ecol, 8(4):293–295. https://doi.org/10.31901/24566608.1997/08.04.13

Ramakrishnan R, Rajendrakumar S, Kothurkar NK (2024) Regional sustainability of the Kattunayakan tribe in Kerala, India through the enhancement of agricultural, livestock, and livelihood options. Agric Syst 217:103929. https://doi.org/10.1016/j.agsy.2024.103929

Rambaree K, Powers MCF, Smith RJ (2019) Ecosocial work and social change in community practice. J Community Pract, 27(1):1–8. https://doi.org/10.1080/10705422.2019.1660516

Rout S (2017) Gendered participation in community forest governance in India. Contemp Soc Sci 13(1):72–84. https://doi.org/10.1080/21582041.2017.1393555

Shein PP, Sukinarhimi P (2021) Taboos as a Social Mechanism Keeping the Human-Nature Balance: Core Values and Practices of Rukai TraditionalEcological Knowledge of Water. Sustain, 14(4):2032. https://doi.org/10.3390/su14042032

Sihlé N (2018) Assessing and adapting rituals that reproduce a collectivity: The large-scale rituals of the Repkong Tantrists in Tibet. Relig Soc,9(1):160–175

Smith DA, Holland MB, Michon A, Herrera F (2017) The hidden layer of indigenous land tenure: informal forest ownership and its implications for forest use and conservation in Panama's largest collective territory. Int For Rev 19(4):478–494. https://doi.org/10.1505/1465548822272356

Snyder K (1997) Elders' authority and women's protest: the Masay ritual and social change among the Iraqw of Tanzania. J R Anthropol Inst. https://doi.org/10.2307/3034767

Soren R (2022) Tribal people, land rights and governance. Asian J Res Soc Sci Human 12(3):105–110. https://doi.org/10.5958/2249-7315.2022.00132.0

Strauch AM, Rurai MT, Almedom AM (2016) Influence of forest management systems on natural resource use and provision of ecosystem services in Tanzania. J Environ Manag 180:35–44. https://doi.org/10.1016/j.jenvman.2016.05.004

Sunil DS (2007) State interventions and natural resource management: a study on social interfaces in a riverine fisheries setting in Kerala, India. Nat Resour Forum 31(1):61–70. https://doi.org/10.1111/j.1477-8947.2007.00128.x

Taba P (2024) Tradition, leadership, and justice in Puroik Village councils: an in-depth examination. Shodhkosh J Visual Perform Arts. https://doi.org/10.29121/shodhkosh.v5.i4.2024.1033

Tseer T, Kuusaana ED, Sulemana M (2023) When does dispute resolution hold? Lessons from three Ghanaian case-studies. Ethnopolitics 24(2):117–138. https://doi.org/10.1080/17449057.2023.2247660

Turyani L, Suharini E, Atmaja HT (2024) Norma Dan Nilai Adat Istiadat Dalam Kehidupan Sehari-Hari Di Masyarakat. Sosial: Jurnal Ilmiah Pendidikan IPS 2(2). https://www.researchgate.net/deref/https%3A%2F%2Fdoi.org%2F10.62383%2Fsosial.v2i2.224?_tp=eyJjb250ZXh0Ijp7ImZpcnN0UGFnZSI6InB1YmxpY2F0aW9uIiwicGFnZSI6InB1YmxpY2F0aW9uIiwicHJldmlvdXNQYWdlIjoic2NpZW50aWZpY0NvbnRyaWJ1dGlvbnMifX0

Vijayalakshmi V (2002) A report on the politics of inclusion: Adivasi women in local governance in Karnataka. Gend Technol Dev 6(2):269–283. https://doi.org/10.1080/09718524.2002.11910046

Wattoo MA, Ali T, Khan MA, Shahbaz B (2010) Stakeholders' role and interactions in context of forestry extension interventions in the Northwest Pakistan. Pak J Agric Sci 47:173–177

Watts S (2008) Institutional constraints on interactive community participation in forest conservation in Mozambique. J Sustain For 26(4):301–327. https://doi.org/10.1080/10549810701879768

Zapf MK (2009) Social work and the environment: understanding people and place. Canadian Scholars' Press, Toronto

Zulu LC (2008) Community forest management in Southern Malawi: solution or part of the problem? Soc Nat Resour 21(8):687–703. https://doi.org/10.1080/08941920802039242

Chapter 8
Forecasting with Fishers: Climate Justice, Local Science, and Early Warning for All

Max Martin, P. Vijayakumar, and S. Abhilash

Abstract As tropical storms intensify across the Arabian Sea in a warming globe, traditional fishing communities along India's southwestern coast face escalating risks to their habitats and livelihoods. Often the changed weather regime involving high wind and waves, uncertain seasons, and severe storm impacts puts their lives at risk too. Weather services available to traditional fishers do not necessarily reflect their local needs and diverse fishing activities at different distances from the shore. In this context, this chapter explores how co-produced, small-grid marine forecasts—developed in collaboration with artisanal fishers, scientists, and social actors—can reshape early warning systems to be more inclusive, precise, and just.

Rooted in field research and participatory practices in the past 8 years, this study situates localised forecasting as both a climate adaptation tool and a justice intervention, finding a place for the fishers in the mainstream risk regime. By integrating climate science, local ecological knowledge, and lived realities of coastal livelihoods, the chapter argues for a more comprehensive and inclusive forecasting and early warning system—one that treats information as a right and risk communication as a shared responsibility. Framed around the principle of early warning for all, this work advocates for forecasting systems that prioritize the most exposed while advancing both social and environmental equity.

M. Martin
Department of Life Sciences, Christ University, Bengaluru, Karnataka, India

Centre for Landscapes, Wildlife and Marine Ecology (CLIME), Christ University, Bengaluru, Karnataka, India
e-mail: max.martin@christuniversity.in

P. Vijayakumar (✉)
Department of Environmental Sciences, University of Kerala, Thiruvananthapuram, Kerala, India

S. Abhilash
Department of Atmospheric Sciences, Cochin University of Science and Technology (CUSAT), Kochi, Kerala, India

Advanced Centre for Atmospheric Radar Research (ACARR), CUSAT, Kochi, Kerala, India

J. K. Joseph, A. P. Pradeepkumar (eds.), *Eco-Social Work for Disaster Risk Reduction*, Disaster Resilience and Green Growth, https://doi.org/10.1007/978-981-95-4885-9_8

Keywords Climate change · Weather forecast for fishers · Community science · Social inclusion · Early warning systems

8.1 Introduction

"In the rainy season, it's dicey." That is all Wilson Franklin said. Those five words captured the monsoon sea—its moods, its risks, the gamble it demands each day. Those words, uttered on the sandy shores of Puthukurichy, a village 20 km north of Thiruvananthapuram city in Kerala state, adjoining rich fishing grounds over underwater reefs close to the shore, moved us. That was a small group of researchers seeking ways to make weather forecasting more accessible and accurate for food producers. When narrated at an international conference in London, those five words moved five eminent professors among the audience, too. Then they advised us to learn more about the local fishing seasons and the risks that the fishers face therein. So those words kicked off our field research, *Forecasting with Fishers*, back in 2016. We were trying to figure out whether the fishers can load the dice in their favour and stay safe.

In the process, we had several scoping and pilot trips and planning sessions at the University of Sussex and later Cochin University of Science and Technology, our academic bases. We talked with fishers—men and women—listened to their stories, learned about the marine weather, examined climate models, produced and issued localised weather forecasts, co-produced weather reports with fishers and forecasters, created feedback loops, and shared our insights with officials, academics and social workers. We traversed a 100 km stretch of the southwestern coast, from Thangassery in Kollam district of Kerala to Enayam in Kanyakumari district of Tamil Nadu. We lived in some of these villages, shared meals with families and boat crews and stories of the sea and their land.

We co-designed forecasts with a few of these fishers—veterans who could read the sea and the sky. Their sons and daughters worked with us. They included social researchers, two aeronautical engineers who produced weather news, a marine biologist who became our field research coordinator and a professional journalist and a computer engineer who managed our online weather radio.

All of us—an interdisciplinary team of 6–20 researchers at different points specialising in social sciences, geography, marine biology and atmospheric sciences—spent time on the coasts, or watching the weather dynamics unfolding over the coastal waters on our computer terminals with the help of satellite imagery, radar data and local observations. Our motto was simple. "Save one life and it's all worth it," as one of our senior advisors put it.

To make forecasts relevant and useful, we needed to produce and share better forecasts at the first mile, where the fisher pushes their boats off to the ocean under a leaden sky. Then the forecasts should reach the last mile, where the fishers go, anywhere from zero to 100 km offshore or even beyond. This chapter explores why such forecasts are inevitable these days, how such forecasts can be made possible, together with the fishers. The chapter tries to encapsulate our research and outreach efforts over the past 8 years and explores the way forward.

First it sets the scene of forecasting at the first mile and looks at how we started this process, step by step, nautical mile by nautical mile. Then it elaborates on the context of fishing, arguably one of the most dangerous professions in the world. It explores the changing storm patterns in the Arabian Sea, an outcome of global warming, and explains the science behind it. Then the chapter narrates how the fishers read the ocean's signs and share their knowledge. As the next step, the chapter narrates the experience of producing forecasts—and marine weather knowledge in general—with fishers and forecasters. It places our work in the context of the global efforts to provide timely early warnings for all. The core argument is that forecasts need to be localised, relevant, timely, seamless and in line with the climate adaptation, disaster risk reduction and sustainable development needs.

8.2 Forecasting on the First Mile

From the air, Thiruvananthapuram, Kerala's capital, looks like a magic tapestry of pristine beaches, emerald mangroves, rocky cliffs, strips of white and golden sand, estuaries, lagoons and lakes. Long strips of barrier islands lie between the shimmering turquoise of the sea and the deep green canopy of coconut groves. Along these shores lives one of the world's largest communities of artisanal fishers—50,000 active seagoing fishers and their families in 42 villages on 78 km (48 miles) of coast. Unlike neighbouring regions, parts of Kanyakumari in the south, Kollam in the north, Thiruvananthapuram has few trawlers and virtually no large-scale operations.

Most fishers here work from small, motorized boats up to 10 m (34 feet) in length, alongside motorized fiberglass canoes and traditional non-motorized catamaran rafts (SIFFS 2017). They live at the edge of society, outside the politics of land and power. Many are poor.

Small-craft fishing here is often a gamble. The tropical sea is rich but fierce, especially during the monsoon. The season begins on 1 June. It lasts till 30 September. That means 122 days of high waves, winds, and sometimes sudden spells of stormy weather that turn the sea rough. Currents run strong. Visibility dips. Weather forecasts do exist, but often many fishers ignore them.

Our first task was to understand the sea and the coast. And listen to the local people. We established contacts with forecasters and state officials. Our research collaborators included the State Disaster Management Authority, India Meteorological Department (IMD) and the Indian National Centre for Ocean Information Services (INCOIS), the national ocean forecasters. We sought ways to produce better forecasts, clear, accurate, and usable. We were committed to co-produce them with the users, who are fishers.

We began our work in Anchuthengu, a barrier island north of the city. In 1690 the East India Company had built a fort here. Now the sea was carving out parts of the village away. Harbour works in a nearby estuary had shifted currents, prevented beach building and contributed to serious erosion of the coast. The local fishers who fished from the harbour wanted to have safety signals for their safe entry into the narrow harbour mouth. Trained designers worked with them to shape a makeshift signal with red and green lamps made from boat accessories that could help them navigate safely.

We helped a local popular science group in Puhtukurichy, a village 10 km north of Anchuthengu set up an experimental online and phone-in marine weather news service called *Radio Monsoon*. They sent marine bulletins in Malayalam (the regional language) over mobile telephony, Facebook and WhatsApp, based on daily IMD and INCOIS feeds. Fishers liked hearing them through local loudspeakers, too. They wanted forecasts along with fish prices and market news. The service continued through our research there and it still continues. Sussex Sustainability Research Programme supported our pilots.

About 35 km to the south of Anchuthengu lies Poonthura. The villages are connected by sea, road, and a canal named after the queen of the erstwhile Travancore kingdom who had it built. Fishers traded and established social connections across these two villages. Poonthura had been torn apart by Cyclone Ockhi in November 2017, a tragic event that killed over 365 fishers from South India at sea without adequate forecasts. We began our work there on the New Year's Eve of 2018, the day the villagers held a night vigil for those departed and gone missing.

Fishers here wanted two things. First, forecasts that reached them where they launched their boats. Loudspeakers. Voice calls. Not text. Second, a system that worked at sea. They often went 25 miles offshore, sometimes more. The winds there were nothing like those near shore. Young fishers already shared news by mobile and wireless up to 10 miles out. They dreamed of a community wireless base station that would relay forecasts to the fishing boats from the village. They built that eventually.

We sat with the fishers, officials, designers and community media practitioners. The fishers in these villages had the same demand—make forecasts local, accurate, and on time. Offshore winds mattered most. The services for big ships were out of reach. But experiments like *Radio Monsoon* tapped into these services and shared them with the fishers.

Government officials, too, were learning. In March, SDMA used loudspeakers to warn of a storm. They banned fishing, a usual practice in the face of frequent storm warnings. The fishers welcomed the alerts, but not too many of them. They wanted precision. Not bans over half the sea. Not too broad warnings.

By May, the monsoon would return. We planned more workshops, more trials. We would map boat tracks by GPS. We would compare them with forecast models and satellite images. Then sit with the fishers and check what matched. Share the results with forecasters. Build better systems.

The aim was simple. Safer seas. Safer lives. Our creed, unspoken but shared, was: *Save one life and it's worth it.*

8.3 Promoting Safe and Sustainable Fishing

We learned and shared with the government forecasting and disaster management agencies a few invaluable lessons in this process: First, fishers are not passive receivers of the weather news. They know their forecasts. They look for them. They make their own, drawing on traditional knowledge and lived experience. They use whatever they can get, from multiple sources, including knowledgeable elders, Internet-savvy youngsters, weather apps, church announcements, lived experiences, real-time marine observations shared over the wireless, websites of the official forecasters, the IMD and INCOIS.

Second, the fishers still often do not get what they need; the detail, the timing, and the language; it is not designed for them. The forecasts speak, sometimes a lot, but not always in a voice they understand. Third, the main character at sea is always the *fish*. Not the forecaster. Not the official who tries to restrict fishing. The central character is the fish that is out there, shining silver, throbbing, splashing under a grey sky. "There's fish. Let's go." That is the spirit.

To fish safely, especially under a grey sky with imminent prospects of rough weather ahead, fishers need forecasts that are local, detailed, robust, and relevant. Something like winds at 30 kilometres an hour (km/h) touching 45 km/h at times, from the southwest, growing stronger after 4 pm, with a swell that might go over 2 m high. Extremely dangerous for small boats of 34 feet and rafts. Winds south of Thiruvananthapuram could be even stronger at 40 km/h.

IMD forecasts of late look like this, but that was not the case during the early phases of our research during 2018. The general tone during the monsoon season was 40 km/h wind on the shores of Kerala; fishers should not go to fish. Air the same text on the following days, then repeatedly over the monsoon season. Fishers ignored those forecasts. They did not trust them. Sometimes they were wrong, as our studies show.

The project found insights into how fishers in the region used weather forecasts and what they needed to improve their safety and preparedness. They found that fishers do use weather forecasts, which were mostly accurate for the period. Fishers also relied on their own knowledge of local conditions, as weather warnings were sometimes perceived as too conservative; fishing activity was driven by the availability of fish to catch. This varied; in Anchuthengu village, fishers went out on 40% of days with advisories against going to sea, but in Poonthura village, this was over 70%, as the Sussex team's 2018 study showed. However, fishers preferred to fish closer to the shore, about 10 km out, during the monsoon season (Martin et al. 2022).

Later we extended the research to seven villages (from north to south): Poonthura, Puthiyathura, Karumkulam, Poovar, Neeordi, and across the state border Neerodi and Vallavilai in Kanyakumari district. Data was collected by tracking the trips of 40 fishing boats from the villages for 122 days in the monsoon of 2019 and then compared with forecasts and wind and wave measurements and 30 years of storm data for the Arabian Sea. This was then set in context by a series of focus groups and interviews with local fishers, discussing forecast needs and responses to weather hazards.

Working with fishers, forecasts and weather alerts were then designed and distributed via mobile phones, radio, and loudspeakers. These activities were supplemented with workshops on risk, where fishers identified risks in local coasts and harbours and suggested a set of measures such as localised forecasts, information on storm tracks, and better signal systems at harbours. The Royal Geographical Society (2021) supported this work.

Understanding the forecasting needs and practices of artisanal fishers has helped provide better targeted weather forecasting to help them manage hazards at sea. Eventually, IMD bulletins became more detailed and the agency acknowledged our contribution. Their top scientists co-authored a research paper with us.

The team has established a template for developing and communicating best practice in co-producing weather knowledge with artisanal fishers, which will enable other communities to benefit from the project's work and findings. This research has led to several additional projects, which will extend activities around the co-production of weather forecasting, as well as testing local forecasts and examining the effect of weather events on coastal erosion (RGS 2025).

Later in a major project supported by UK Research and Innovation, we spoke to more than 300 men and 30 women in the fishing villages of Thiruvananthapuram. We observed their work. We had regular conversations. We logged fishing trips. We learned how the men decided when to go and how women managed the household finances, even when men could not fish.

Fishers read the sky, the wind, and the sea. They hear the forecast, but do not necessarily trust it.

Still, fishers want the science. They demand better science. They want it close, local, in their tongue. They want to help make it, and to help send it out. They want daily forecasts they can act on—not only to stay home, but to prepare, to plan, to fish in spite of danger.

Our partners at CUSAT and Sussex University built a three-tier forecast. Seven days for the Arabian Sea. Three days for Kerala waters. Daily, high-resolution forecasts for fishing grounds near the shore. Models checked against reanalysis data, checked against the fishers' own eyes. The Weather Research Forecasting tool gave significant results, especially for moderate winds. Less so for the light or the heavy. But still better than what they had before.

We worked on the message too. A mobile app in Malayalam. A website. Audio bulletins through Radio Monsoon, a community station. Loudspeakers on the sand. We trained young fishers to read graphics, speak weather, and spread the word. They gave feedback, and we adjusted. Simple charts, clearer signals. Public screens to show what the sea was doing. Fishers used the tools. They asked for more. They demanded to be part of the work—production, communication, and management. Their trust grew when the forecasts were theirs too (Osella 2024).

8.4 The Most Dangerous Profession in the World

Fishing is one of the most dangerous professions in the world, even in some of the Western countries with stringent labour safety standards. Day-to-day fishing is a risky affair as such. Monsoon brings dangerous sea states, that is, living with risks over a period of 122 days from June to September. Then overlapping with the start (March–June) and just after (October–December), the monsoon comes the storm seasons of the North Indian Ocean. Stretching across the Arabian Sea (5–25° N, 55–75° E) to the west and the Bay of Bengal (5–22° N, 80–100° E) to the east of peninsular India, this part of the ocean remains warmer than before, spawning and intensifying more storms.

First, the day-to-day risks involved in fishing are many. Local fishing in our study area is largely artisanal. There are older fishers who use shore seine nets. Shore seining involves casting a net in a wide arc through shallow coastal waters to encircle a shoal of fish. The net is set by a country boat and then hauled ashore by two groups of men standing in long lines on either side, their rhythm guided by old work songs. Others cast nets from estuaries and river mouths. Some free dive alone for shells and conch near rocky reefs. Many go out in *kattumaram* rafts or small fibreglass boats with outboard motors, working within ten nautical miles. Others venture as far as 60 nautical miles in 30-foot boats with 9–40 horsepower engines.

There are hook-and-line fishers operating near the coastal reefs and in deep water. Some travel in pairs or small fleets for boat seine work, some spend a week at sea in 38-foot boats. Further south, in northern Kanyakumari across the Kerala-Tamil Nadu border, deep-sea fishers chase sharks in 48-foot steel boats fitted with machines, staying out 4–5 weeks at a time. Together, they form a diverse and resilient community. For them, the coast, offshore, and deep sea are all places of work.

As they navigate the sea, they blend information from formal weather sources with their own observations, memories, and ancestral knowledge.

To stay safe, these fishers watch many things: local and offshore winds, sudden gusts, wave height and direction, surface and deep-sea currents. And they do all this in small, fragile boats. There is no captain's training, no ground control, no standard operating procedure. Often, not even a life jacket. They go head-on into the sea. It is a gamble, but a calculated one. The fishers know the limits of their boats. During the monsoon, it is the wind, waves, fog, and lightning that bring danger. Of these, small-boat fishers fear the wind most. Studies show they stop work when wind speeds reach 25 knots (about 29 miles/h). On the Beaufort scale, that is a "strong breeze." Locally, fishers say they avoid going out in 30- to 34-foot boats when winds are about to reach that level.

Wind stirs up dangerous waves. The surface waves you see at the beach are rhythmic and familiar. But storms bring stronger, rougher waves, even rogue giants. A boat can ride waves up to a certain size. But it matters how the wave hits. Front or back, the boat can stay stable. From the side, it rolls and can flip. Big waves also rise when shallow waters slow them, or when winds blow against the current. In some estuaries, waves become massive due to river outflow or the funnelling effect of narrowing shores.

During Cyclone Ockhi, fishers at sea reported waves taller than two-storey buildings. Such waves outlive the storms that made them. They can travel thousands of kilometres over several days, hiding their strength and size beneath the ocean surface as long-period swells. Even when they look gentle as and roll on for days and days, they turn violent near shore.

In April 2018, a cyclone far away in the southern Indian Ocean sent waves that struck Kerala's coast 5 days later—wrecking boats and homes. More followed in the years ahead, some powerful enough to destroy entire villages. Swells here are gaining height, as our ongoing studies show, even as storms grow fiercer.

Monsoon days are the most dangerous. Between 2011 and 2016, there were 643 fishing accidents on the southwestern edges of India, comprising Kollam, Thiruvananthapuram, and Kanyakumari districts. Seventy-five percent involved small boats with outboard engines, the most prevalent fishing boats in this region of artisanal fishers. One in five were struck by waves while crossing breakers. Capsizing made up 13%, port/harbour accidents 7% (SIFFS 2017). These boats usually launch from the sandy beaches. But in rough seas, they use harbours. And when harbours are badly designed, that brings new risks.

Monsoon is a season of good catch. The fish come close to shore. Cold water rises from the deep, rich with life. Rivers bring silt. The winds churn the sea. The rains cool it. The fishers say so, and they are right. In these months there is no mechanized fishing. It is the breeding season. Only the small boats go out. The old fishers have the markets to themselves. Still, the work is hard. The catch is never sure. Fish spoil fast. There is little ice, little storage. Income is uncertain, the season is short, and the product is quick to rot (Martin et al. 2022). So, when the fish come close, no one lets the chance slip.

At Muthalappozhi harbour in northern Thiruvananthapuram (near Anchuthengu), rogue waves have caused deaths and disasters (Shaji 2023). According to media reports, at least 70 fishermen have lost their lives in these waters—caught in high waves, undercurrents, and swirling eddies, their boats capsizing or crashing against rock-lined shores.

Then are swells created by a faraway storm. Before the monsoon, long-period swells generated by storms and sustained winds in the Southern Indian Ocean or South Atlantic travel north to the southwest coast of India (Remya et al. 2016; Kumar and Sivakrishnan 2023). These swells take 4–6 days to cross nearly 10,000 km before reaching shore (INCOIS 2024). Their approach shifts with the season: from the southwest during the non-monsoon months, and from the west-southwest during the monsoon (Shanas et al. 2024). Unlike short, choppy surface waves driven by local winds, these rolling swells are smooth, powerful, and carry greater energy. When they break, they pound the shoreline, damaging habitats and crushing coastal infrastructure.

8.5 Storm Warning

Warming seas, strong winds, and clustered cyclones are rewriting the Arabian Sea's weather script. The Indian Ocean is warming up; its heat fuels storms. And the winds that once sheared them apart have weakened. The balance has tipped. Tauktae (2021) and Biparjoy (2023) in the Arabian Sea; Amphan (2020) and Mekunu (2018) in the Bay of Bengal and western Arabian Sea, respectively—all Extremely Severe Cyclonic Storms (ESCS: sustained winds of 167–221 km/h 90–119 knots). Most formed in May, just before the southwest monsoon sets in.

North Indian Ocean storms strike in two distinct phases. One comes before the monsoon, between March and June. The other follows it, from October to December. It is during this post-monsoon season that the most dangerous pattern is emerging. In October 2014, Cyclone Nilofar became the first Extremely Severe Cyclonic Storm ever recorded in the post-monsoon Arabian Sea. Its winds roared past 110 knots, a signal that the game had changed. The following year, Chapala and Megh—two more ESCSs—formed back-to-back in the same season, something that had never happened before.

Then came 2019. Five named storms stirred the Arabian Sea. Two of them—Maha and Kyarr—were monsters. Maha was an ESCS. Kyarr was even more intense—a Super Cyclonic Storm (wind above 222 km/h or 120 knots). For a brief moment, the two giants roamed the sea together. That had not happened in the satellite era. The changes are not just about how many storms form; it is about how long they last. The total time under Very Severe Cyclonic Storm (VSCS) conditions, winds between 118 and 166 km/h or 64 and 89 knots, has tripled. Even milder cyclones, those with sustained winds from 34 knots upward, now linger longer than they once did (Abhiram Nirmal et al. 2023).

One powerful example was Cyclone Ockhi—a Very Severe Cyclonic Storm (VSCS) in 2017. Born in the Bay of Bengal, it swept westward into the Arabian Sea, feeding on anomalously high sea surface temperatures (SSTs) and unusually warm subsurface waters. That deep reservoir of heat was key to Ockhi's rapid intensification. The storm did not follow a conventional path. It intensified quickly, crossed basins, and defied forecast models.

Ockhi was not just a freak storm; it was a clear and loud signal of the new storm regime taking shape. A 2023 paper we co-authored for *Nature Scientific Reports* looked closely—layer by layer, year by year, from 1979 onward. The ocean's structure has changed. And with it, the sky (Abhiram Nirmal et al. 2023).

Traditionally, cyclone studies relied on Sea Surface Temperature (SST) as a key metric. But today, researchers look deeper into the sea's hidden energy stores. Ocean Heat Content (OHC) measures the total thermal energy stored in the upper ocean, while Tropical Cyclone Heat Potential (TCHP) captures the energy available down to the 26 °C isotherm, a line of constant temperature found about 164–328 feet (50–100 m) beneath the surface. This deep warmth is no passive backdrop. It is active fuel, immense energy that cyclones can tap into, especially when surface winds weaken and atmospheric instability grows.

The *Scientific Reports* study showed that the Eastern Arabian Sea has seen a sharp increase in cyclone activity since 1979. Of late, more storms have formed off Kerala and Karnataka. These are often stronger, longer-lasting, and harder to predict. The study traced their intensification to increased subsurface warming, higher mid-tropospheric humidity, and rising thermal instability, especially during the pre-monsoon (MAMJ) and post-monsoon (OND) periods.

Moisture now saturates the mid-level atmosphere, especially post-monsoon. Instability grows. Vertical lift strengthens. Add deep heat and moisture; storms follow. Storms do not just appear. They take shape through a process called cyclogenesis, when deep ocean heat, moist air, and shifting winds align across sea and sky.

The findings are clear: the clustering is no accident. It is a signal. A warning. By clustering, we mean the growing tendency of cyclones to form in close succession—both in time and geography—particularly in the Eastern Arabian Sea. As the *Scientific Reports* study showed, storms are no longer isolated events. They are emerging in bursts. This pattern is linked to deep subsurface warming, higher mid-tropospheric humidity, and rising atmospheric instability, all of which create fertile conditions for rapid storm formation and intensification.

8.6 A Sea Warming Ahead of the Curve

The Indian Ocean is warming faster than the global average (Roxy et al. 2014). Especially the Arabian Sea. Hotspots spread where currents shift and winds falter. Heat rises. So does the sea. And with it, the storms. They are not what they used to be.

The northern Indian Ocean, bordered by nations with millions of people, now sees stronger cyclones and more violent rain. Not rare outliers. A pattern. A new rhythm in the sea. Extremely Severe Cyclonic Storms are more frequent. Storms form faster. Track longer. Intensify rapidly. In just decades, their power and number have grown.

There is a warm stretch of sea, the Indo-Pacific Warm Pool, that extends from the eastern Indian Ocean to the western Pacific. A hot zone: over 28 °C year-round. It fuels storms. Over time, this warm pool has doubled in size. It shifts monsoons. Drives rainfall.

As the ocean warms, the changes are not just visible in the air or on the surface. The very structure of the sea is shifting. Thermal expansion, water swelling with heat, raises sea levels. Melting ice from glaciers and ice sheets adds more. The result: coasts erode, beaches vanish, and backwaters flood.

Amid all this, the ocean does not lie still. It breathes—rising, falling, pulsing with heat and wind, turning furious occasionally.

One of the ocean's deepest breaths is the Madden–Julian Oscillation, or MJO, a slow-moving pulse of cloud, wind, and rain that travels around the equator every 30–60 days. This pulse begins over the warm waters of the Indian Ocean, rolling eastward along the equator through Indonesia and the Pacific. As it passes, it stirs the air and deepens the sea's breath. Storms often follow in its wake.

Other pulses shape the ocean, atmosphere dialogue too, most notably El Niño and La Niña, together known as ENSO: the El Niño–Southern Oscillation. Born in the Pacific, ENSO behaves like a see-saw. When El Niño dominates, warm water builds in the eastern Pacific, trade winds weaken, and atmospheric pressure flips. Rainfall drifts away from Asia. The Indian monsoon stumbles. Drought follows. When La Niña rises, the see-saw tips the other way, cool waters spread eastward, trade winds regain strength, and cyclone activity in the Indian Ocean tends to increase.

This climate oscillation tilts the odds of storms, droughts, and floods across continents. In El Niño years, the Indian monsoon often weakens. Crops fail. Reservoirs dry. In La Niña years, rainfall may return, but with it, the risk of intense storms rises, especially in the Indian Ocean. Alongside these see-saws of wind and warmth, another giant engine drives the weather's great choreography: the Hadley Cell.

The Hadley Cell, a vast, tropical circulation loop where warm air rises near the equator, moves poleward high in the atmosphere, sinks in the subtropics, and returns as surface trade winds. For centuries, this engine has helped regulate where it rains and where it stays dry. But as the oceans warm, the Hadley Cell is expanding, its pushing its sinking zones farther toward the poles. As this belt shifts, the zones where cyclones form also drift northward, bringing stronger storms into new latitudes, including the Arabian Sea.

As the Hadley Cell expands, its sinking zones shift farther from the equator, displacing rainfall belts and altering wind patterns. In South Asia, this can delay the onset of the southwest monsoon, shorten its duration, or weaken its reach in traditionally rain-fed areas.

Some regions may get more intense but short-lived rainfall, while others may experience longer dry spells, disrupting agriculture, water supplies, and livelihoods. The result is an increasingly erratic monsoon, less tied to the old calendar and harder to predict with precision.

The Walker Circulation, a vast loop of rising air over the western Pacific and sinking air over the eastern Pacific, moves east to west, driving the Pacific trade winds and shaping rainfall across the tropics. When this circulation weakens or reverses, warm water shifts eastward, triggering El Niño. The balance breaks, the pressure flips, and weather patterns ripple outward across oceans and continents. The ocean speaks, and the sky answers, sometimes halfway around the world.

These are the teleconnections, vast, interlinked climate engines that no longer follow the old rhythms. In the Arabian Sea, their overlapping signals are rewriting the storm script. In its eastern stretch, a new storm nexus has emerged. Deep ocean heat, widening atmospheric cells, and seasonal lulls in wind shear now align during the monsoon, March to June, and again from October to December, spawning cyclones that cluster, form close in time and space, intensify quickly, and defy old expectations.

8.7 Reading the Ocean's Signs

For artisanal fishers, last-mile forecasts are the holy grail. But at sea, the "last mile" varies widely. For deep-sea tuna fishers, it can be 500 nautical miles offshore, or even farther. For crews on 30-foot fibreglass boats, it is around 50 miles. For shore-seine fishers, working boats lashed with rope, it is just five. And for free divers harvesting shellfish near the coast, it is virtually zero.

Each group watches the sky, reads the sea, and shares whatever updates they can, through mobile phones, short-range wireless, or word of mouth. All live in fear of storms, especially during the volatile pre- and post-monsoon months. Once, the Arabian Sea was calmer than the Bay of Bengal. Not anymore.

Staying clear of storms, chasing unseen fish, and conserving ice and fuel require delicate judgment.

That is how they sense risk. Wilson's two-word dictum—"It's dicey"—is more than a philosophy. It reflects the science behind a fisher's decision-making. It is like what Hemingway said: the deeper meaning should be hidden, like the submerged part of an iceberg. Wilson was sketching a logic of judgment that is broad, time-bound, and precise. A fisher needs a deep, fast sense of what is happening: wind, waves, tides, currents. They often work with limited data and must check updates several times a day. And many things, like the winds at sea, cannot be seen from shore. The sea may look calm. But if the engine is not strong enough, one sudden wave can hit the boat hard.

Because weather and sea conditions shift quickly, fishers have developed ways to read them clearly and act early. They draw on three kinds of knowledge.

- First, official forecasts, alerts from the IMD, INCOIS, the fisheries department, police, disaster authorities, and district offices.
- Second, digital and crowd-sourced updates, from private and international apps, online platforms, amateur weather watchers, and researchers who share updates via wireless networks and social media.
- Third, traditional and lived knowledge—what elders have passed down, what they have learned through years at sea, and what they exchange with peers. This includes real-time signals: wind shifts, cloud patterns, wave sounds—shared by wireless, mobile, word of mouth, even shouts and whoops across boats and rafts.

While at sea, they constantly cross-check and update this knowledge using mobile phones, radios, and instinct, stitching together forecasts, memory, and moment. Studies show that fishers combine multiple sources of information remarkably effectively. Wilson and his friends acknowledged that forecasts are especially useful during the monsoon but also pointed out that more studies are needed to understand how fishers actually interpret and use them in practice.

Fishers try to make the best of the information they have. For example, in Puthiyathura, fishers used to gather outside the church after early mass. There, a man named Augustine would read the sea forecast aloud. He would explain wind speeds, directions, wave heights, and warnings, breaking down the technical data into useful guidance. He had learned this from years of experience, training sessions, asking questions, and trial and error. When Augustine was not around, others would step in. Younger fishers often brought forecasts on their phones. But not everyone had a smartphone. Not everyone could read the charts. So, they listened. They talked it through.

In some villages, forecasts come over loudspeakers from the harbour office. In others, they arrive via WhatsApp groups run by local officials. But none of it works without trust. Fishers do not blindly follow a forecast. They test it. Compare it. Ask if it matches the clouds, the wind, the colour and smell of the sea. Their experience.

Some ignore forecasts completely. They say they change too often, or do not reflect the local weather. Some prefer to rely on instinct. But more and more fishers are starting to use forecasts. A few now read wave models. Some can tell what is coming just by the feel of the wind. Others simply want to know one thing: Is it safe or not?

Forecasts help reduce risk. But they do not eliminate it.

The sea is inherently volatile. A forecast offers probabilities, not certainties. That is why fishers do not rely on data alone. They read the water: the tilt of the swell, the direction of the wind, the sudden spiral of seabirds. These are signals honed over generations, often more immediate, more trusted than any satellite map.

And then they decide.

8.8 Making Forecasts Together

Our programme, *Forecasting with Fishers*, had a simple aim: to keep fishers safe. It sought to make the lives of South India's small-boat fishers more secure in a time of storms. The sea had turned harsher. The weather shifted fast. Forecasts came late, or not at all. Many went to sea blind. Some never returned.

This work brought fishers together with forecasters and government men. They sat at the same table. They spoke the same words. Out of this came forecasts that meant something to the men on the water. Not numbers, not jargon, but clear words they could trust. This cuts risk. It saved lives and boats. It helped men judge when to fish and when to stay ashore.

The plan did not end on this coast. It shaped a set of tools, an "action template," that could be used in other villages, other seas. A way to make forecasts fit the people who needed them.

We, a wide-ranging team, worked on it: anthropologists, geographers, weathermen, ocean men, experts in machines, and media. They walked the beaches, listened to the fishers, and watched how they fished. We studied the sea with satellites and buoys. They built new ways to send warnings—onshore, offshore, by phone, by voice, by radio. We tested what worked in three fishing grounds: Kanyakumari, Thiruvananthapuram, and Kollam. This coast is crowded with fishers, boats, and nets. It is one of the richest fishing stretches in Asia. It was also one of the riskiest. The impact was clear. Men got the warnings they needed.

8.9 Early Warnings and Forecasts for All

Not all fishers are the same. A man on a raft close to shore has different needs from the crew of a fiberglass boat that sails 70 km out. The raft fisher must know if the sea will turn rough within 5 km, and in the next 4 h. The boat crew must know the sea state for the day, far from land. Yet the warnings they receive are the same. Broad. Generic. Issued for the "Kerala coast."

Such warnings are blunt tools. They tell every fisher to stay ashore when the seas are heavy. But that is not possible. A man's family must earn. When the warnings are too wide, they become useless. They strip fishers of choice. They strip them of the right to decide their own risk.

A warning system that does not speak to the first mile is an empty promise. It is a right denied. To know the risks of your own sea is not a privilege. It is survival.

In the villages, men and women know their lives are exposed. They speak of accidents, broken boats, deaths in the surf. They ask for information they can trust. Instead, they receive forecasts that stretch too far, cover too much, and often exaggerate the danger. Over-forecasting breeds doubt. When the warnings do not match what the sea delivers, trust breaks. Then, when the real danger comes, men ignore the alerts.

Localized Early Warning Early Action can cut the impact of climate hazards. It works by enabling well-planned, locally led actions before a hazard strikes. For fishers, this matters. Rough monsoon seas often overturn small craft. Swell waves wash away habitats and flood low-lying homes. These are seasonal, frequent, and predictable. Which means there is merit in folding EWEA into regular marine weather forecasts, along with warning messages. A seasonal or sub-seasonal advisory can prevent repeated distress from high waves and swells. It can break the cycle of predictable shocks, short-term responses, and costly recovery. Over the long run, clearer guidance on storm trends can help avert humanitarian disasters (UNDRR 2022).

Such warnings do more than save lives. They inform planning for climate adaptation and sustainable development (Martin et al. 2021). Yet, gaps remain. A generic warning for the Kerala coast asks all fishers to abandon fishing. This is impractical and unnecessary, unless the event is a full storm. Blanket warnings reduce useful choices on the last mile.

There is scope for sharper communication. Fishers remember Cyclone Ockhi in 2017 as the turning point. They see hazards changing and demand better, more accurate, and more local forecasts. But forecasters still fall short. Advisories tell fishers to stay ashore. Most ignore this. Instead, they want precise, timely forecasts to weigh risks and plan trips. The problem is local relevance. Different fishers need different scales of information. For shore seine crews, the wave breaking near the beach matters. For raft fishers, it is the sea state within 10 km. For fiberglass boats and ring seiners, the concern is within 75 km. Offshore trips need a wider marine forecast.

Studies confirm this gap. Warnings often arrive with multiple response options, but without enough detail to guide good anticipatory action (Lentz and Maxwell 2022). Probability is rarely clear. People guess weather outcomes and often guess wrong. Over-forecasting also erodes trust. When high winds are forecast but fail to materialize, credibility falls. To reverse this, users must be placed at the front end of planning. Work from likely hazards and needed actions back to information design.

Research shows that relevant, localised forecasts save lives and livelihoods. They prevent undue restrictions and build trust. Solutions exist. They include better observation systems, stronger models, and collaboration across governance levels and communities. Fishers want detail. They want to know the sea and sky near their shores, and the state of distant waters where their craft may reach. Their first need is safety for today's trip. But early warning also helps them plan for the next.

8.10 Way Forward

The goal is simple. Give fishers forecasts they can act on. Local. Timely. Probabilistic. Actionable. That is the only measure of success. The three-tier framework. A 7-day outlook. A 3-day coastal forecast. A daily high-resolution product for the fishing grounds. Each tier must match the boat and the risk. Rafts need nowcasts at a 5-km, 4-h scale. Offshore boats need daily maps at 50–75 km. Measure success by use. Count the men who change their plans because of it. The system must start at the shore. Put sensors in the water and on the coast—buoys, gauges, anemometers, logbooks. Link them with satellite eyes and radar. Feed this into the models. Correct the bias. Run ensembles. Check performance by error, bias, and skill. But also, by the wind thresholds that matter to a boat at sea: safe, limit, stop.

Forecasts must speak the language of risk. Not pressure fields or vector maps, but surf at the harbour mouth, swell on the bar, waves on the fishing ground. Co-design the thresholds with the fishers. Test the words, the icons, the audio clips. Use loudspeakers, radios, WhatsApp voice, and apps if they help. Iterate until the message is clear and trusted.

Verification must be open. Benchmarks against ERA5 and local records. Scores made public. But numbers alone are not enough. Ask who checked the forecast. Who stayed ashore? Who came back safe?

Readiness matters as much as data. Map the safe channels and shelters. Run drills tied to forecast triggers. Link the warnings to action—boats pulled ashore, markets warned, and harbours cleared. Count the volunteers who show up. Count the drills completed.

Institutions must talk. IMD, INCOIS, SDMAs, fisheries, and universities. Set rules for alerts, duties, and feedback. Let local centres be the bridge for training and trust. Finance must follow: cash for days lost, insurance tied to verified triggers, safety nets for when the sea is closed.

Research must go on. Swells from far storms. Seasonal predictability in the eastern Arabian Sea. Trade-offs between income and safety Trials of radio and low-bandwidth tools. Each project must deliver something that reduces loss or raises adoption.

Monitor. Evaluate. Build dashboards that track skill, uptake, accidents, and gear lost. Use what you learn to refine the system and scale it to the next coast.

And above all, hold to ethics. Forecasts are right. No village, caste, gender, or crew should be left outside the signal. A forecast that widens inequality is a failure, no matter its score.

The synthesis is plain. Science must cut uncertainty. Models must serve need. Co-production and trust must close the loop.

Forecasting with Fishers is not a lab trial. It is a fight for rights. To deny them a timely, usable forecast is to leave them open to harm. To secure that right is to build stewardship—science, state, and community working as one. That is the first mile of safety at sea.

Acknowledgement This research was funded by the UK Research and Innovation, Sussex Sustainability Research Programme, Royal Geographical Society and the Global Disaster Preparedness Center of the American Red Cross. We gratefully acknowledge weather news services by Radio Monsoon, based on forecasts of the India Meteorological Department and Indian National Centre for Ocean Information Services, with tech support from Gram Vaani. We thank Aaditeshwar Seth (Indian Institute of Technology Delhi), Harikumar R (INCOIS) and their colleagues for their shared expertise and Filippo Osella, Dominic Kniveton and Joseph Alcamo (University of Sussex) for advice and guidance. We are grateful to our field and lab colleagues who made our work possible and the local fishers who made it relevant.

A seed money grant from Christ University, Bangalore, made recent field research and data analysis possible.

References

Abhiram Nirmal CS, Abhilash S, Martin M et al (2023) Changes in the thermodynamical profiles of the subsurface ocean and atmosphere induce cyclones to congregate over the Eastern Arabian Sea. Sci Rep 13:15776. https://doi.org/10.1038/s41598-023-42642-9

Indian National Centre for Ocean Information Services (2024) Swell surge alert to Kerala, South Tamil Nadu, Lakshadweep, Andaman and Nicobar Islands, Karnataka, Goa, Maharashtra, Gujarat, Northern Andhra Pradesh, Odisha and West Bengal. Press release, May 5, 2024. https://incois.gov.in/documents/PressRelease_SwellSurgeEvent_May-05-2024.pdf

Kumar VS, Sivakrishnan KK (2023) Ocean surface wave dynamics off the southern tip of Indian mainland.Dyn Atmos Ocean, 104:101411

Lentz E, Maxwell DG (2022) Anticipating, mitigating, and responding to crises: how do information problems constrain action? https://doi.org/10.2139/ssrn.4043459

Martin M, Todd M, Osella F, Kniveton D, Rowhani P, Taylor O, Alcamo J (2021) Forecasting for food producers. SSRP Policy Brief 7. University of Sussex and Institute of Development Studies (IDS). https://www.sussex.ac.uk/webteam/gateway/file.php?name=ssrp-policy-brief%2D%2D-forecasting-for-food-producers-final-.pdf&site=492

Martin M, Abhilash S, Pattathil V, Harikumar R, Niyas NT, Balakrishnan Nair TM, Grover Y, Osella F (2022) Should I stay or should I go: South Indian artisanal fishers' precarious livelihoods and their engagement with categorical ocean forecasts. Weather Clim Soc 14:113–129. https://doi.org/10.1175/WCAS-D-20-0044.1

Osella F (2024) Forecasting with fishers data for 3 coastal villages of Thiruvananthapuram District, India, 2020–2022. Data collection. UK Data Service, Colchester. https://doi.org/10.5255/UKDA-SN-856463

Remya PG, Vishnu S, Praveen Kumar B, Balakrishnan Nair TM, Rohith B (2016) Teleconnection between the North Indian Ocean high swell events and meteorological conditions over the Southern Indian Ocean. J Geophys Res Oceans 121:2169–9275. https://doi.org/10.1002/2016JC011723

Roxy MK, Kapoor R, Terray P, Masson S (2014) The curious case of Indian Ocean warming. J Clim 27:8501–8509. https://doi.org/10.1175/JCLI-D-14-00471.s1

Royal Geographical Society (2021) Improving safety and sustainability in food production by co-producing weather forecasts. https://www.rgs.org/impact/fishingforecasts

Shaji KA (2023) Blood-thirsty Muthalapozhi in Kerala gobbles up lives as fishermen keep sailing into the 'mouth of death'. The South First. https://thesouthfirst.com/kerala/on-muthalapozhi-the-mouth-of-death-for-fish-workers-far-south-on-kerala-coast/

Shanas PR, Sanil Kumar V, Sivakrishnan KK, Mandal S, Yuvaraj S (2024) Impact of local and remote winds on the surface waves in the coastal waters of eastern Arabian Sea. Ocean Eng 295:116714. https://doi.org/10.1016/j.oceaneng.2024.116714

SIFFS (South Indian Federation of Fishermen Societies) (2017) Sea safety incidents on the lower south west coast of India, Thiruvananthapuram. SIFFS Doc., 40 pp

United Nations Office for Disaster Risk Reduction (UNDRR) (2022) Global status of multi-hazard early warning systems: target G. UNDRR, WMO, UNDRR Bonn Office. https://iddrr.undrr.org/publication/global-status-multi-hazard-early-warning-systems-target-g

Chapter 9
Technological Interventions in Eco-Social Work

Arun Anil and Shahnaz Basheer

Abstract Eco-social work prioritises social and environmental justice as key tenets and highlights the connection between ecological health and human well-being. Disasters disproportionately impact vulnerable communities, women, children, the elderly, and marginalised groups because they exacerbate poverty, relocation, and inequality in the backdrop of worsening climate change and environmental degradation. Conventional social work techniques alone are not enough to address these issues; intentional, moral integration of technology is necessary to strengthen communities and foster resilience.

This chapter examines the important technical tools used in disaster preparedness, response, recovery, and resilience building, including Geographic Information Systems (GIS), participatory early warning systems, social media platforms, mobile applications, drones, and satellite imaging. It illustrates how these interventions improve situational awareness, ensure fair resource distribution, and elevate underrepresented perspectives through case studies and real-world applications. In addition, it critically examines issues like algorithmic bias, data justice, the digital divide, and the dangers of "techno-solutionism," emphasising the value of community-centred, culturally sensitive, and rights-based strategies.

Based on sustainability, participation, human rights, and social justice, the chapter suggests an Integrated Eco-Social Disaster Technology Framework to handle these challenges. As two cornerstones of fair and sustainable disaster management, this approach emphasises proactive risk reduction and equitable recovery. In the end, the chapter argues that, when used in accordance with eco-social principles, technology can serve as a catalyst for empowerment, equity, and long-term resilience rather than merely being an efficient tool.

Keywords Eco-social work · Technological interventions · Disaster management · Environmental and social justice · Resilience

A. Anil (✉) · S. Basheer
Loyola College of Social Sciences (Autonomous), Thiruvananthapuram, Kerala, India
e-mail: arunanil012@gmail.com; shahnazbasheer.geo@gmail.com

J. K. Joseph, A. P. Pradeepkumar (eds.), *Eco-Social Work for Disaster Risk Reduction*, Disaster Resilience and Green Growth, https://doi.org/10.1007/978-981-95-4885-9_9

9.1 Introduction

In a time of climate change and environmental degradation, the intensity and frequency of disasters are increasing and threatening human societies. The field of eco-social work identifies and works towards social well-being and environmental health. "Social work has to develop a theoretical framework to incorporate the ecological environment because man and environment are crucial and mutually dependent components" (Saini and Singh 2023). It underscores that the vulnerable and marginalised groups, like women, children, the elderly, people with disabilities, and the poor, are facing the impacts of these disaster events. These vulnerable communities face loss of livelihood, forced migration, trauma and also increase the existing social inequalities. Social work needs to engage with environmental issues like climate change and environmental degradation through interdisciplinary collaboration (Coates and Gray 2011). Addressing these challenges requires an approach rooted in social and environmental justice, ensuring that disaster management efforts are not only practical but also fair and equitable.

This chapter argues that a deliberate and ethical technology integration is crucial for advancing social and environmental justice in disaster management. Therefore, this chapter will explore various technological tools such as Geographic Information Systems (GIS), social media platforms, and early warning systems and their potential to empower communities, ensure equitable resource distribution, and build sustainable resilience.

9.1.1 The Eco-Social Work Perspective

Human well-being is fundamentally interconnected with ecosystem functioning, as ecosystems provide essential provisioning, regulating, cultural, and supporting services (Butler et al. 2003). Well-being encompasses basic material needs, health, social relations, security, and freedom of choice, all affected by ecosystem changes (Butler et al. 2003). Eco-social work is a specialized field that recognizes the deep and inseparable link between human well-being and the planet's health. It operates on the core principle of interconnectedness, asserting that environmental degradation directly impacts social welfare, and social problems are often linked to environmental issues. This approach moves beyond traditional social work to consider the broader ecosystem as a key part of human life.

Environmental justice and social justice are deeply interconnected concepts that address the equitable distribution of environmental benefits and burdens across society. Environmental justice represents an essential social justice component, as environmental regulations and policies significantly impact quality of life, health, and access to natural resources (Millner 2011). The twin pillars of Social Justice and

Environmental Justice are at the heart of eco-social work. Social justice ensures equitable rights and opportunities for all people, focusing on protecting the most vulnerable from harm. Environmental justice, in turn, addresses the unfair distribution of environmental risks and benefits, advocating for the right of all people to a healthy and safe environment, regardless of their social or economic status.

The Sustainable Development Goals (SDGs) represent a comprehensive framework for global transformation, building upon the foundational concept of sustainable development defined by the Brundtland Commission as "development that meets the needs of the present without compromising the ability of future generations to meet their own needs" (Clark and Wu 2016). This is not just about environmental protection; it's about meeting the needs of the present without compromising the ability of future generations to meet their own needs. Therefore, eco-social work aims for socially fair and environmentally sound solutions, ensuring that disaster management efforts contribute to building resilient communities and ecosystems for the long term.

9.1.2 Introducing Technology as a Tool

Technology can serve as a powerful tool for empowerment and justice in disaster contexts when guided by appropriate principles, though significant challenges remain. Technology has great potential to promote development and progress, but its implementation must be guided by ethical considerations and social justice commitments (Ebua 2023). While traditional social work methods are essential, the scale and complexity of modern disasters require innovative tools. This chapter proposes that when guided by the core principles of eco-social work, technology can be a powerful instrument for promoting empowerment and justice in disaster contexts. Rather than viewing technology as a neutral force, it should be seen as a dynamic tool that can either reinforce existing inequalities or actively work to dismantle them. The key is to ensure that its application is ethical, equitable, and community-centred.

GIS applications in social work remain underutilized despite their potential for understanding environmental contexts and supporting qualitative research through participatory mapping approaches (Teixeira 2018). By leveraging technologies such as GIS, mobile communication tools, and data analytics, eco-social workers can enhance early warning systems, facilitate equitable resource distribution, and amplify the voices of marginalized populations. Therefore, technology's practical and ethical use becomes crucial in advancing the eco-social work goals of fairness and sustainability in disaster risk reduction.

9.2 A Framework for Technology in Eco-Social Disaster Work

9.2.1 Principles for Practice

By placing human well-being within a broader ecological and social context, eco-social work emphasises sustainability, equity, and justice as interconnected goals (Dominelli 2012). These guidelines must be followed when integrating technology into disaster contexts to ensure that short-term interventions are successful and to promote long-term resilience and systemic change. The four core dimensions of sustainability, participation, human rights, and social justice, along with holistic approaches that place an emphasis on collaboration and systemic change, are used in this section to present a framework for technology in eco-social disaster work.

The fundamental principles ensure that technology use in disaster response and recovery activities maximises beneficial social effects while also being in line with ethical principles. Additionally, by implementing trauma-informed computing principles, technology integration must prioritise user safety and foster trust, especially for vulnerable populations that may have experienced trauma (Chen et al. 2022).

Integrating the ideas of responsible innovation and human-centred design, the Integrated Eco-Social Disaster Technology Framework (Fig. 9.1) integrates the Sustainable Development Goals and the Sendai Framework for Disaster Risk Reduction (UNDP 2024; Abdelaziz and Mesbah 2021). This framework is organised around two fundamental pillars to address disaster management issues

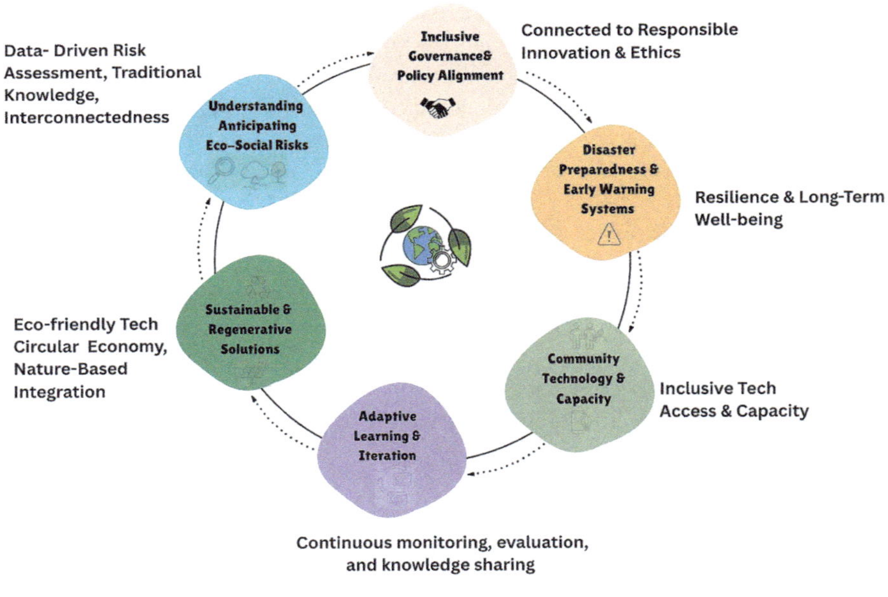

Fig. 9.1 The Integrated Eco-social Disaster Technology Framework (Source: Authors)

holistically: Equitable and Effective Response and Recovery, and Proactive Risk Reduction and Resilience Building (Peters and Peters 2021).

To promote an anticipatory rather than a reactive disaster management paradigm, the first pillar, Proactive Risk Reduction and Resilience Building, strongly emphasizes using technical advancements to identify, evaluate, and mitigate disaster risks before their manifestation. Together with participatory design approaches that guarantee the co-creation of technological solutions with impacted populations, this entails utilizing sophisticated data analytics, remote sensing, and GIS to comprehend intricate vulnerabilities within ecological systems and human communities (Paudel et al. 2024; Vasileiou et al. 2022). The people these technologies are intended to assist should also be involved in their design (Asian Development Bank 2021). The objective of utilizing technology to build more robust, socially equitable, and ecologically sustainable systems from the ground up is to transcend a reactive approach. This strategic integration of technical capability with community participation is being implemented to move beyond traditional risk assessment and toward a more nuanced understanding of systemic vulnerabilities and the promotion of sustainable preventative interventions (Rahman et al. 2018). This proactive approach is further supported by the framework's dedication to ongoing monitoring and adaptive management, which enables dynamic modifications to risk reduction tactics in response to real-time data and changing environmental conditions. Furthermore, by combining traditional ecological knowledge with technology innovations, the framework promotes creating ecologically sound and culturally relevant hybrid solutions that increase community resilience through various strategies.

Ensuring that technological interventions promote rather than obstruct equitable outcomes, the second pillar, Equitable and Effective Response and Recovery, identifies the ethical and inclusive deployment of technology during and after disasters. This entails tackling issues of digital literacy and access while also leveraging technology such as crowdsourced data and mobile communication to distribute needs-based help, transparently, and quickly (UNDP 2024; Abdelaziz and Mesbah 2021). According to the paradigm, technology cannot make already-existing disparities worse. Instead, it ought to be a tool for community empowerment, empowering them to take charge of their recovery and rebuild more effectively (UNDRR 2015). Instead of only offering temporary solutions, this strategy ensures that technology promotes ecological health and long-term community well-being.

9.3 Technology Across the Disaster Management Cycle

9.3.1 Mitigation and Preparedness

Technology is essential for understanding and reducing risks during the pre-disaster mitigation and preparedness phase. GIS technology is valuable for social workers and planners to understand client communities and environmental contexts through spatial analysis and mapping capabilities (Teixeira 2018). By overlaying data on physical hazards, such as flood zones and earthquake faults, with social data, like poverty rates, age demographics, and disability status, GIS can show which communities are most at risk. This information helps shape specific mitigation strategies and resource distribution, ensuring that preparedness efforts focus on the most marginalized populations. This approach promotes environmental and social justice.

Beyond complex mapping, technology supports community-level preparedness with accessible tools. Early warning systems that use simple mobile phone technology, like SMS alerts and automated voice calls, can deliver timely and critical information directly to at-risk individuals. This is especially important in areas with limited internet access. Community radio, digital forums, and social media platforms can also help share preparedness knowledge, local disaster plans, and safety protocols. This fosters a more informed and ready community.

Finally, digital technologies are crucial in building community resilience by enabling collaborative information sharing and strengthening social networks during crises (Belblidia 2010; Mehan 2023). These technologies facilitate virtual communication, remote access to resources, and community engagement, proving particularly valuable during emergencies and disasters (Mehan 2023). By encouraging communities to engage in their preparedness actively, technology shifts the focus from top-down directives to a more collaborative and empowering approach, which aligns well with the core principles of eco-social work.

9.3.2 Response

Technology saves lives and helps reach those in need during the immediate response phase. Technology, particularly social media platforms and crowdsourcing tools, has revolutionized disaster response by enabling real-time communication and citizen-driven data collection. During emergencies, affected individuals use platforms to share locations, report urgent needs, and provide damage information (Imran 2017; Landwehr and Carley 2014). This creates a dynamic information flow that contrasts traditional top-down communication methods (Poblet et al. 2014). The Kerala floods demonstrated how communities can rapidly mobilize resources through technology-facilitated crowdsourcing during rescue and relief phases (Ajay 2019). Social media enables people to broadcast food, water, and shelter needs, while disaster relief organizations extract actionable information to understand evolving situations (Imran 2017; Landwehr and Carley 2014). These platforms generate georeferenced real-time information on critical events, helping humanitarian organizations gain situational awareness and deploy resources more effectively (Poblet et al. 2014; Imran 2017). Simple, locally aligned technologies often prove more effective than advanced equipment in disaster management contexts (Ajay 2019).

Technology is increasingly transforming humanitarian aid distribution and response mechanisms. Digital payment systems, mobile banking, and electronic vouchers enable direct financial transfers to affected populations, improving user-friendliness and accountability while reducing traditional delivery delays (Smith et al. 2011). This method helps people buy what they need and prevents corruption in conventional aid programmes. By using these technologies, social workers can ensure that relief efforts are more open and accountable. This approach directly supports the most vulnerable populations and upholds the principles of social justice. However, most aid funding still flows to international organizations rather than local actors, potentially undermining social justice principles despite humanitarian guidelines emphasizing participatory, rights-based approaches (Haanyika et al. 2024).

Technology adoption in social work and humanitarian contexts presents opportunities and challenges for serving vulnerable populations, requiring careful consideration of ethical standards, privacy, and boundary issues (Branson and Miller 2020). Innovative technologies like drones for mapping and assessment are being implemented across humanitarian organizations. However, their introduction raises essential questions about moral responsibilities toward vulnerable populations and the risk of prioritizing technological solutions over addressing structural problems (Wang 2019).

9.3.3 Recovery and Resilience

Technology, big data and remote sensing technologies are crucial in disaster resilience and recovery efforts. Big data technologies enhance disaster management by improving the speed and effectiveness of linkages between disaster information and systemic response, with key tools including remote sensing imagery, social media data, crowdsourced data, GIS, and mobile metadata (Sarker et al. 2020). These technologies support all phases of disaster management through training, early warning systems, emergency evacuation, and relief distribution (Sarker et al. 2020).

Satellite imagery and unmanned aerial vehicles (UAVs) are particularly compelling for rapid damage assessment, enabling emergency responders to allocate resources and facilitate recovery efforts efficiently (Krichen et al. 2025). Advanced analytics, including convolutional neural networks and semantic segmentation, can automate the analysis of satellite and UAV imagery for rapid damage mapping and classification (Krichen et al. 2025). Integrating crowdsourced data with remote sensing inputs further improves situational awareness and damage reporting capabilities (Krichen et al. 2025).

In rural and remote communities, targeted e-health strategies can increase accessibility and quality of mental health services, though technology supplements rather than replaces face-to-face interaction in recovery processes (Burmeister and Marks 2016). Digital health and telemental health approaches are increasingly important for addressing global disasters' rising mental health impacts, providing care to previously underserved populations (Augusterfer et al. 2018).

9.4 Key Technological Interventions in Practice

9.4.1 *GIS for Justice*

"A Geographic Information System (GIS) is defined as a computer system that is used for capturing, storing, managing, querying, analysing and displaying geospatial data" (Chang 2019). GIS is a type of spatial cartography and a technological subspecialty that makes it possible to store, manipulate, and visually display data (Dunlop et al. 2021). By combining information on social aspects like poverty rates, access to healthcare, and the whereabouts of marginalized groups with data on physical dangers (like flood zones), GIS is used in eco-social work to produce "vulnerability maps" (Chang et al. 2021). To guide targeted mitigation initiatives and ensure equitable resource allocation, this multi-layered approach assists in identifying communities that are most at risk (Chen et al. 2021). For instance, VGI and open socioeconomic data were combined with machine learning during an intense downpour in Zhengzhou, China, to identify vulnerable individuals, their needs, and possible safe locations with shelter supplies (Chen et al. 2021).

9.4.2 *Community-Centred Early Warning Systems (EWS)*

By utilizing accessible technologies like social media, mobile applications, and SMS warnings, EWS goes beyond conventional sirens. Messages are culturally relevant, understood, trusted, and accessible to everyone, including those with low literacy or disabilities, when these warning systems are co-designed with communities (Hermans et al. 2022). To improve disaster risk reduction, this participatory method is essential for combining scientific and local knowledge (Hermans et al. 2022). One example is the creation of a localized landslide early warning system for Rohingya refugees in Bangladesh. This system incorporates a dynamic web-based alert system, rainfall thresholds, landslide inventory and susceptibility maps (Ahmed et al. 2020).

9.4.3 *Mobile Technology for Response and Recovery*

During and after disasters, smartphones and mobile applications are essential for quick data collection, direct money transfers, and effective service delivery. While digital voucher systems and mobile banking offer impacted individuals and families immediate cash support, social workers can use mobile apps to assess requirements quickly. By enabling disaster survivors to buy what they need and avoiding potential bottlenecks and corruption in conventional relief delivery channels, this strategy promotes autonomy and dignity for individuals (UNDP 2024). In disaster risk mitigation, digital tools such as social media and participatory mapping make it easier for stakeholders to engage and share knowledge (Sechi et al. 2023).

9.4.4 *Drones and Satellite Imagery for Assessment*

Particularly in remote or hazardous areas, unmanned aerial vehicles (UAVs or drones) and satellite imagery provide a thorough aerial perspective for quickly and safely evaluating damage to houses, infrastructure, and natural habitats (Majumder et al. 2022; Ebrahim 2025). Long-term recovery strategies, rescue priorities, and aid distribution depend on this knowledge (Hanifa et al. 2022). For instance, to assess landslide vulnerability and provide early warning systems for the Rohingya refugee crisis, geoinformation techniques were used to monitor widespread deforestation and hill cutting in Cox's Bazar, Bangladesh (Ahmed et al. 2020). However, these powerful surveillance technologies raise important ethical questions about privacy, data protection, and community consent. To support an eco-social approach, we need to address these concerns using the framework's principles of human rights and participation. This will ensure that technology benefits the community without compromising its dignity.

9.5 Critical Challenges and Ethical Considerations

9.5.1 *The Digital Divide*

The first significant challenge is the digital divide, which refers to the gap between those with access to technology and those without access (Hollimon et al. 2025; Matenga-Ikihele Te Whānau-Ā-ApanuiNiue et al. 2023). In a disaster, this gap can become a matter of life and death, as access to information and resources is often tied to technology (Taylor and Junod 2024). Technological solutions, if not carefully planned, can inadvertently leave the most vulnerable people, such as people with low incomes, the elderly, or those in rural areas, even further behind (Wang et al. 2023).

From an eco-social work perspective, addressing this divide requires a commitment to equitable access (Matenga-Ikihele Te Whānau-Ā-ApanuiNiue et al. 2023). This means not only providing technology but also ensuring that people have the necessary skills to use it, and that the chosen technologies are appropriate for the local context (Wang et al. 2023). For instance, relying on a complex smartphone app may be ineffective in areas with low literacy or limited internet connectivity (Marshall et al. 2023). Therefore, technological interventions must be paired with human-centred strategies to bridge the gap and ensure no one is excluded. Failing to address the digital divide directly weakens the fundamental social principle of justice.

9.5.2 Data Justice and Privacy

In a disaster, large amounts of sensitive data, including personal information, health records, and location data, are collected from survivors (Kreutzer et al. 2025). This raises critical questions of data justice and privacy. Who owns this data? How is it stored? And who has access to it? This information could be misused without clear policies, leading to surveillance, discrimination, or exploitation.

A core principle of eco-social work is protecting individuals' and communities' rights and dignity (Visave 2024). This extends to their digital lives and they must ensure that survivors have control over their own data and give informed consent for its use (Squire et al. 2024). Organizations must be transparent about their data collection practices and implement strong security measures to protect sensitive information from unauthorized access (Kreutzer et al. 2025). The goal is to use data to help people, not to control or exploit them.

9.5.3 Algorithmic Bias

As we increasingly use Artificial Intelligence (AI) and algorithms to make decisions, such as prioritizing aid distribution, we must be aware of the risk of algorithmic bias (Gevaert et al. 2021). People create algorithms and can unintentionally reflect the biases of their creators or the data they are trained on (Rocca et al. 2023). For instance, an algorithm designed to identify those most in need of aid might overlook historically underrepresented communities in official data or accidentally favour specific demographies over others (Kondmann and Zhu 2021).

This kind of technological bias can lead to unequal and unjust outcomes, directly contradicting the goals of eco-social work (Visave 2025). To prevent this, developers and social workers must work together to create transparent, fair, and regularly audited algorithms (Rocca et al. 2023). The data used to train these systems must be diverse and representative, and there must always be human oversight to ensure that technology serves people, not discriminating against them (Rocca et al. 2023).

9.5.4 Avoiding "Techno-Solutionism"

Finally, a key ethical challenge is avoiding "techno-solutionism," the belief that technology alone can solve deep-rooted social and environmental problems (Sætra and Selinger 2023). While technology is a powerful tool, it is not a magic wand (Allen 2024). Poverty, inequality, and environmental injustice are complex and require human-led, community-driven solutions (Wang and Altanbulag 2022). Over-reliance on technology can lead to a dehumanizing approach that ignores the social and emotional needs of disaster survivors (Beduschi 2022).

From an eco-social work perspective, technology must be seen as a support, not a replacement, for human connection and community leadership (Wu and Greig 2022). The most effective interventions combine technological efficiency with human empathy, face-to-face interaction, and the active participation of local communities (Munawar et al. 2022). The true power of technology lies in its ability to amplify human efforts and empower communities to lead their own recovery, not to take that power away from them (Mitra et al. 2023).

It is emphasized that technology must support, not replace, human connection and community leadership. True eco-social work focuses on human connection and empowerment. This shows the principle of participation.

9.6 Summary of Key Arguments

This chapter has examined the critical role of technology in eco-social work for disaster risk reduction. Technological tools, such as GIS and mobile applications, are not just for technical experts; they can be strong tools for promoting social and environmental justice. The study observed how technology can help identify vulnerable populations, deliver aid more fairly, and empower communities to strengthen their resilience. However, the study also pointed out the serious risks that these tools can pose. Issues such as the digital divide, data injustice, and the threats of "techno-solutionism" can worsen existing inequalities if technology is not used carefully and thoughtfully. The main point is clear: the effectiveness of technology in eco-social work relies not on the tool itself but on how the fundamental principles of justice, participation, and human dignity guide it.

9.6.1 Recommendations for Eco-Social Workers

9.6.1.1 Prioritize Participation

The best technological solutions are those made with communities, not just for them. Eco-social workers should always involve community members in the selection, design and implementation of technological tools. This approach ensures that the technology fits the culture, addresses real needs, and enables local people to manage their resilience.

9.6.1.2 Build Digital Literacy

Providing technology is not enough; we need to work with communities to help them use these tools safely and effectively. This includes training on how to find reliable information, safeguard personal data, and use digital platforms to advocate for their rights. Digital literacy empowers people and helps close the gap between having access to technology and using it to create meaningful change.

9.6.1.3 Advocate for Digital Equity

Eco-social workers ought to be ardent supporters of laws that guarantee equitable access to technology. This entails advocating for inclusive digital infrastructure, accessible devices, and reasonably priced internet, particularly in isolated and underserved places. Building truly inclusive disaster resilience requires bridging the digital divide, which is a social justice issue.

9.6.1.4 Future Directions

The field of eco-social work and technology intersection is one that is changing quickly. Looking ahead, new technologies like the Internet of Things (IoT) for real-time environmental monitoring and artificial intelligence (AI) for forecasting the effects of disasters on certain social groups present intriguing new opportunities. But these developments also raise new issues. Critical topics like how to create community-owned data models that safeguard the privacy and sovereignty of marginalised groups should be the main focus of future studies. What are the long-term psychological and social effects of disaster recovery technology use? How can we produce more culturally grounded and successful solutions by fusing modern technology with indigenous or local knowledge? In order to direct future ethical technology use and responsible innovation, it will be crucial to investigate these issues.

9.7 Concluding Thought

While integrating technology into eco-social activities, we must never lose sight of our ultimate goal. The objective is not only to develop more intelligent systems or effective tools, but also to employ those tools to make the world a more compassionate, just, and equitable place. This calls for a holistic strategy that recognises the complexity of development, including social advancement, economic expansion, and environmental sustainability. However, with technological breakthroughs mostly focused on environmental domains like biodiversity preservation and climate change mitigation, the social dimension of sustainability is frequently overlooked in both theory and practice. Technology in eco-social work must serve humanity. Its success should be judged not by its efficiency but by its dedication to the dignity and well-being of the people and the planet we serve.

References

Abdelaziz A, Mesbah S (2021) A review of disaster management frameworks. J Manag Inform Decision Sci 24(S5):1–10

Ahmed B, Rahman MS, Sammonds P, Islam R, Uddin K (2020) Application of geospatial technologies in developing a dynamic landslide early warning system in a humanitarian context: the Rohingya refugee crisis in Cox's bazar, Bangladesh. Geomat Nat Haz Risk 11(1):446–468. https://doi.org/10.1080/19475705.2020.1730988

Ajay A (2019) Role of technology in responding to disasters: insights from the great deluge in Kerala. Curr Sci 116(6):913–918

Allen HJ (2024) Fintech and techno-solutionism. Available at SSRN 4686469

Asian Development Bank (2021) Leveraging technology and innovation for disaster risk management and financing. Retrieved from https://www.adb.org/sites/default/files/publication/664501/tech-innovation-disaster-mgt-financing.pdf

Augusterfer EF, Mollica RF, Lavelle J (2018) Leveraging technology in post-disaster settings: the role of digital health/telemental health. Curr Psychiatry Rep 20(10). https://doi.org/10.1007/s11920-018-0953-4

Beduschi A (2022) Harnessing the potential of artificial intelligence for humanitarian action: opportunities and risks. International Review of the Red Cross 104(919):1149–1169

Belblidia MS (2010) Building community resilience through social networking sites. International Journal of Information Systems for Crisis Response and Management 2(1):24–36. https://doi.org/10.4018/jiscrm.2010120403

Bill R, Blankenbach J, Breunig M, Haunert JH, Heipke C, Herle S et al (2022) Geospatial information research: state of the art, case studies and future perspectives. PFG J Photogramm Remote Sensing Geoinform Sci 90(4):349–389

Branson DC, Miller KA (2020) Harnessing technology in the social work discipline: moving forward ethically. J Sociol Social Work 8(2):15–25. https://doi.org/10.15640/jssw.v8n2a2

Burmeister OK, Marks E (2016) Rural and remote communities, technology and mental health recovery. J Inf Commun Ethics Soc 14:170–181

Butler CD, Chambers R, Chopra KR, Dasgupta P, Duraiappah AK, Kumar P, McMichael AJ, Wells J (2003) Ecosystems and human well-being. Island Press, Washington, DC

Chang KT (2019) Introduction to geographic information systems. McGraw-Hill Education, New York

Chang H, Pallathadka A, Sauer J, Grimm NB, Zimmerman R, Cheng C et al (2021) Assessment of urban flood vulnerability using the social-ecological-technological systems framework in six US cities. Sustain Cities Soc 68:102786

Chen Y, Ye Z, Liu H, Chen R, Liu Z, Liu H (2021) A GIS-based approach for flood risk zoning by combining social vulnerability and flood susceptibility: a case study of Nanjing, China. Int J Environ Res Public Health 18(21):11597

Chen JX, McDonald A, Zou Y, Tseng E, Roundy KA, Tamersoy A et al (2022) Trauma-informed computing: towards safer technology experiences for all. In: Proceedings of the 2022 CHI conference on human factors in computing systems. ACM, pp 1–20

Clark H, Wu H (2016) The sustainable development goals: 17 goals to transform our world. Furthering the work of the United Nations. United Nations

Coates J, Gray M (2011) The environment and social work: an overview and introduction. Int J Soc Welf 21(3):230–238. https://doi.org/10.1111/j.1468-2397.2011.00851.x

Dominelli L (2012) Green social work: from environmental crises to environmental justice. Polity Press, Malden

Dunlop JM, Chechak D, Hamby W, Holosko MJ (2021) Social work and technology: using geographic information systems to leverage community development responses to hate crimes. J Technol Hum Serv 40(3):201–229. https://doi.org/10.1080/15228835.2021.1931635

Ebrahim MAB (2025) Advancements in surveying and GIS for disaster resilience: protecting public health in vulnerable regions. Biomed J Sci Tech Res 60:009530. https://doi.org/10.26717/BJSTR.2025.60.009530

Ebua EJ (2023) Investigating the potential of technology to promote development and the ethical and social implications of technological innovation in the context of development. Open Access Library J 10(04):1–23

Gevaert CM, Carman M, Rosman B, Georgiadou Y, Soden R (2021) Fairness and accountability of AI in disaster risk management: opportunities and challenges. Patterns 2(11):100363

Haanyika Z, Trauernicht N, Martinetti A, Lutters E (2024) Is humanitarian aid funding fair and square? Correlating social justice to humanitarian aid utilisation. In: 2024 IEEE Global Humanitarian Technology Conference (GHTC). IEEE, pp 481–488

Hanifa NR, Gunawan E, Firmansyah S, Faizal L, Retnowati DA, Pradipta GC, Imran I, Lassa JA (2022) Unmanned Aerial Vehicles for geospatial mapping of damage assessment: a case study of the 2021 *Mw* 6.2 Mamuju-Majene, Indonesia, earthquake during the coronavirus disease 2019 (COVID-19) pandemic. Remote Sens Appl 28:100830. https://doi.org/10.1016/j.rsase.2022.100830

Hermans TD, Šakić Trogrlić R, van den Homberg MJ, Bailon H, Sarku R, Mosurska A (2022) Exploring the integration of local and scientific knowledge in early warning systems for disaster risk reduction: a review. Nat Hazards 114(2):1125–1152

Hollimon LA, Taylor KV, Fiegenbaum R, Carrasco M, Garchitorena Gomez L, Chung D, Seixas AA (2025) Redefining and solving the digital divide and exclusion to improve healthcare: going beyond access to include availability, adequacy, acceptability, and affordability. Front Digital Health 7:1508686

Imran M (2017) Enabling rapid disaster response using artificial intelligence and social media. Qatar Computing Research Institute

Kondmann L, Zhu XX (2021) Under the radar – auditing fairness in ML for humanitarian mapping. arXiv preprint arXiv:2108.02137

Kreutzer T, Orbinski J, Appel L et al (2025) Ethical implications related to processing of personal data and artificial intelligence in humanitarian crises: a scoping review. BMC Med Ethics 26:49. https://doi.org/10.1186/s12910-025-01189-2

Krichen M, Abdalzaher MS, Shaaban MF, Aburukba R (2025) Using aerial videos for natural disaster damage assessment. In: 2025 7th international youth conference on Radio Electronics, Electrical and Power Engineering (REEPE), pp 1–6

Landwehr P, Carley KM (2014) Social media in disaster relief. In: Data mining and knowledge discovery for big data. Studies in big data. Springer, Berlin, Heidelberg

Majumder R, Jana S, Menon PP, Ghose D, Prusty NM, Mukherjee B, Ghosh A (2022) Cyclone preparedness, rescue operations and damage assessment using UAVs. In: 5th world congress on disaster management: volume I. Routledge, London, pp 235–244

Marshall A, Wilson C-A, Dale A (2023) New pathways to crisis resilience: solutions for improved digital connectivity and capability in rural Australia. Media International Australia 189(1):24–42. https://doi.org/10.1177/1329878X231183292

Matenga-Ikihele Te Whānau-Ā-ApanuiNiue A, Fa'alau F, Dobson R, Fa'alili-Fidow J, Roberts M, Taufa S, Tuesday R, Whitakker R, McCool J (2023) Navigating digital inclusion and the digital vā among Niue mamatua through the provision of mobile phones during COVID-19. AlterNative Nga Pae Maramatanga (Organ) 19(1):145–154. https://doi.org/10.1177/11771801221148343

Mehan A (2023) The role of digital technologies in building resilient communities. Bhumi Planning Res J 10(1):33–40. https://doi.org/10.4038/bhumi.v10i1.92

Millner F (2011) Access to environmental justice. Deakin Law Review 16:189–207

Mitra S, Raskin JP, Pansera M (2023) Role of ICT innovation in perpetuating the myth of techno-solutionism. arXiv preprint arXiv:2309.12355

Munawar HS, Mojtahedi M, Hammad AW, Kouzani A, Mahmud MP (2022) Disruptive technologies as a solution for disaster risk management: a review. Sci Total Environ 806:151351

Paudel S, Dambacher JM, Neugart R, Tenzing J (2024) Mainstreaming ecosystem-based approaches into disaster risk reduction policies: a comparative study of Nepal, India, and Bangladesh. J Environ Policy Plan Adv. https://doi.org/10.1080/1523908X.2024.2361732

Peters K, Peters LER (2021) Terra incognita: the contribution of disaster risk reduction in unpacking the sustainability–peace nexus. Sustain Sci 16:1173–1184. https://doi.org/10.1007/s11625-021-00944-9

Poblet M, García-Cuesta E, Casanovas P (2014) Crowdsourcing tools for disaster management: a review of platforms and methods. Lect Notes Comput Sci 8929:261–274. https://doi.org/10.1007/978-3-662-45960-7_19

Rahman A, Sakurai A, Munadi K (2018) The analysis of the development of the Smong story on the 1907 and 2004 Indian Ocean tsunamis strengthens the Simeulue Island community's resilience. Int J Disaster Risk Reduction 29:13–23. https://doi.org/10.1016/j.ijdrr.2017.07.015

Rocca R, Tamagnone N, Fekih S, Contla X, Rekabsaz N (2023) Natural language processing for humanitarian action: opportunities, challenges, and the path toward humanitarian NLP. Front Big Data 6:1082787

Sætra HS, Selinger E (2023) Technological remedies for social problems: defining and demarcating techno-fixes and techno-solutionism, 19 Sept 2023. Available at SSRN: https://ssrn.com/abstract=4576687 or https://doi.org/10.2139/ssrn.4576687

Saini S, Singh NP (2023) Role of social work in sustainable development: towards eco-social practice. Shodh Kosh J Visual Perform Arts 4(1). https://doi.org/10.29121/shodhkosh.v4.i1.2023.2440

Sarker MN, Peng Y, Yiran C, Shouse RC (2020) Disaster resilience through big data: way to environmental sustainability. Int J Disaster Risk Reduction 51:101769

Sechi GJ, Hendriks E, Pregnolato M (2023) Digitalization in disaster risk reduction: the use of smartphones to enhance the safety of informal settlements in Iringa. Tanzania Int J Disaster Risk Sci 14:171–182. https://doi.org/10.1007/s13753-023-00483-0

Smith G, Macauslan I, Butters S, Trommé M (2011) New technologies in cash transfer programming and humanitarian assistance. Concern Worldwide, Oxford Policy Management, Partnership for Research in International Affairs and Development, The Cash Learning Partnership

Squire V, Olayinka A, Jones B, Kuyang L, de Albuquerque JP (2024) Engaging data literacies in displacement. J Human Affairs 5(3):12–22. https://doi.org/10.7227/JHA.113. ISSN 2515-6411

Taylor C, Junod AN (2024) To protect disabled people during climate disasters, we need to invest more in rural connectivity. Retrieved from: https://www.urban.org/urban-wire/protect-disabled-people-during-climate-disasters-we-need-invest-more-rural-connectivity

Teixeira S (2018) Qualitative geographic information systems (GIS): an untapped research approach for social work. Qual Soc Work 17:23–29

UNDP (2024) Innovation in disaster management. United Nations Development Programme. https://www.undp.org/sites/g/files/zskgke326/files/2024-03/innovation_in_disaster_management_web_final_compressed.pdf

UNDRR (2015) Sendai framework for disaster risk reduction 2015–2030. United Nations Office for Disaster Risk Reduction. https://www.preventionweb.net/files/43291_sendaiframeworkfordrren.pdf

Vasileiou K, Barnett J, Fraser DS (2022) Integrating local and scientific knowledge in disaster risk reduction: a systematic review of motivations, processes, and outcomes. Int J Disaster Risk Reduction 81:103255. https://doi.org/10.1016/j.ijdrr.2022.103255

Visave J (2024) AI in emergency management: ethical considerations and challenges. J Emerg Manag Disaster Commun 5(01):165–183

Visave J (2025) Transparency in AI for emergency management: building trust and accountability. AI Ethics 5:3967–3980. https://doi.org/10.1007/s43681-025-00692-x

Wang N (2019) "A success story that can be sold"?: a case study of humanitarian use of drones. In: IEEE International Symposium on Technology and Society (ISTAS). IEEE, pp 1–6. https://doi.org/10.1109/ISTAS48451.2019.8938015

Wang P, Altanbulag A (2022) A concern for eco-social sustainability: background, concept, values, and perspectives of eco-social work. Cogent Soc Sci 8(1):2093035. https://doi.org/10.1080/23311886.2022.2093035

Wang Y, Hao H, Kang S (2023) Disaster risk communication and digital vulnerability among subsidized housing residents. Natural Hazards Center mitigation matters research report series, report 12. Natural Hazards Center, University of Colorado Boulder. https://hazards.colorado.edu/mitigation-matters-report/disaster-risk-communication-and-digital-vulnerability-among-subsidized-housing-residents

Wu H, Greig M (2022) Adaptability, interdisciplinarity, engageability: critical reflections on green social work teaching and training. Healthcare (Basel) 10(7):1245. https://doi.org/10.3390/healthcare10071245

Chapter 10
Mental Health and Well-Being in Post-disaster Settings: An Eco-Social Work Approach

S. P. Sreya and Neenu Mathews

Abstract This chapter looks into the critical importance of addressing mental health and well-being in post-disaster settings through an eco-social work lens. It examines how disasters disrupt not only the physical environment but also the social and psychological fabric of affected communities. The chapter outlines an integrated approach that combines trauma-informed care with environmental restoration, recognizing the therapeutic role of reconnecting individuals with their natural environments.

Keywords Mental health · Well-being · Post-disaster recovery · Eco-social work · Trauma-informed care

10.1 Introduction

Disasters are complex events that disrupt the normal functioning of communities and societies, often resulting in significant human, economic, environmental, and psychological consequences. A disaster is defined as a serious disruption of the functioning of a community or a society at any scale due to hazardous events interacting with conditions of exposure, vulnerability and capacity, leading to one or more of the following: human, material, economic and environmental losses and impacts (UNDRR 2017). As per the Disaster Management Act 2005, a disaster is defined as "a catastrophe, mishap, calamity, or grave occurrence in any area, arising from natural or man-made cause, or by accident or negligence which results in

S. P. Sreya
Department of MSW Disaster Management, Loyola College of Social Sciences, Thiruvananthapuram, Kerala, India

N. Mathews (✉)
Inter-University Centre for Disability Studies, Mahatma Gandhi University, Kottayam, Kerala, India

J. K. Joseph, A. P. Pradeepkumar (eds.), *Eco-Social Work for Disaster Risk Reduction*, Disaster Resilience and Green Growth, https://doi.org/10.1007/978-981-95-4885-9_10

substantial loss of life or human suffering or damage to, and destruction of property, or damage to, or degradation of, environment, and is of such a nature or magnitude as to be beyond the coping capacity of the community of the affected area." India's geoclimatic conditions and socioeconomic factors render it highly vulnerable to a spectrum of natural hazards, including earthquakes, cyclones, tsunamis, floods, droughts, landslides, avalanches, and forest fires. Twenty-seven of the nation's 36 states and Union Territories are prone to natural disasters. Of the 7516 km of coastline, 5700 km are vulnerable to cyclones and tsunamis; 68% of the cultivable land is susceptible to drought; hilly regions are vulnerable to avalanches and landslides; 15% of the landmass is susceptible to landslides; 58.6% of the landmass is susceptible to moderate to very high intensity earthquakes; and 12% of the land is susceptible to floods and river erosion. Urban flooding puts 5161 Urban Local Bodies at risk. Further hazards include fires, industrial mishaps, and other man-made catastrophes involving chemical, biological, and radioactive elements. These have highlighted the need for bolstering mitigation, readiness, and response strategies.

10.2 Background

Rising vulnerabilities associated with changing demographic and socioeconomic conditions, unplanned urbanisation, development in geographically vulnerable areas, environmental degradation, climate change, geological hazards, epidemics, and pandemics further increase India's risk of disasters. These factors undoubtedly play a part in India's economy, population, and sustainable growth, which are at grave risk from disasters (National Institute of Disaster Management 2014). A disaster affects a community in various ways. First, regardless of the number of lives lost, disasters threaten injury or death to a large population. Second, they impact social processes, leading to a loss of resources for the community and a disruption of social networks and services. Thirdly, they have adverse repercussions, specifically measurable health outcomes, both physical and mental, for the individuals impacted.

The declining mortality toll over the decades is clearly demonstrated in Figs. 10.1 and 10.2, reflecting significant improvements in health outcomes and reductions in death rates over time. This suggests that the non-fatal effects of disasters are increasingly significant, particularly their adverse impact on health. Disasters have been shown to precipitate somatic manifestations such as headaches, fatigue, abdominal pain, and shortness of breath. In addition, several adverse impacts on mental health have been linked to disaster exposure. Despite the fact that most people recover successfully from disasters, a significant percentage suffer from psychological impairment, and a lesser percentage will eventually develop mental illnesses.

It is typical and indicative of normal coping and adaptability to be able to carry on with daily activities following a stressful incident. This tendency is now referred to in literature as "resilience." In contrast to those who have a longer period of dysfunction and a slower return to baseline performance, resilient people often

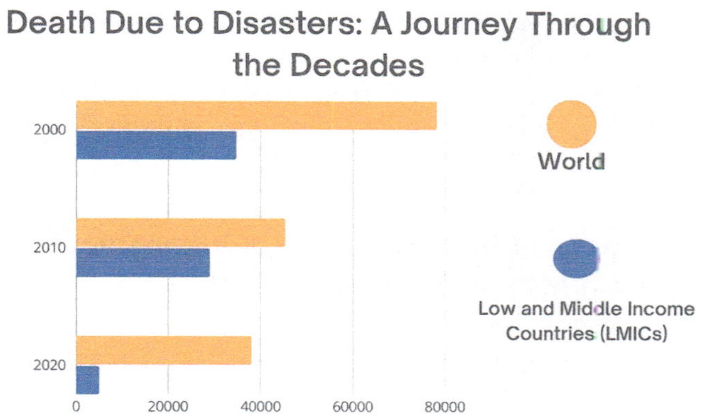

Fig. 10.1 Disaster-related mortality over decades. (Source: Adopted from Public EM-DAT Platform)

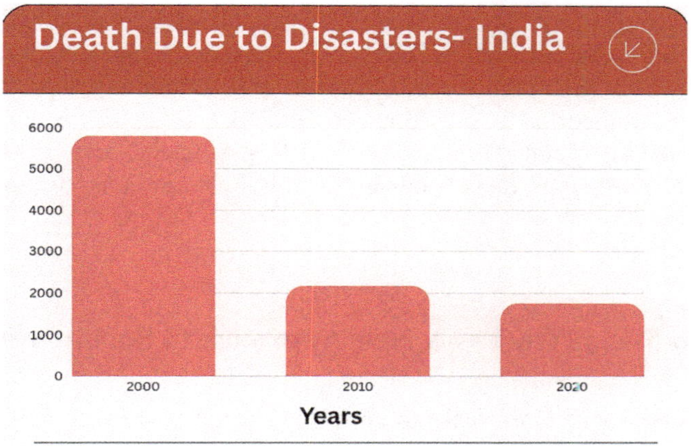

Fig. 10.2 Disaster-related mortality over decades in India. (Source: Adopted from Public EM-DAT Platform)

experience distress for a brief period of time and swiftly recover to pre-disaster levels of functioning. Post-disaster psychopathology is commonly characterized by post-traumatic stress disorder (PTSD), major depressive disorder, substance use disorders, and a range of additional psychological symptoms. These symptoms may include nonspecific psychological distress, heightened perceived stress, suicidal ideation, and feelings of remorse (Goldmann and Galea 2014). This highlights the necessity of urgent mental health intervention, particularly in the post-disaster phase. This chapter primarily looks into the post-disaster impacts on the mental health and well-being of the affected communities. It also delves into the eco-social approach of social work in promoting mental well-being in a post-disaster setting. For this review, the keywords such as "Eco-social work," "mental health and well-being in the disaster settings," "green social work," "eco-social perspective of psychology" were utilized to derive a number of articles from scholarly databases such as PubMed, Google Scholar and Research Gate. In addition, many journals were also consulted.

10.3 Convergence of Multiple Fields in Imparting Mental Health Support in Post-disaster Setting

Uncertainty, panic, and urgency are characteristics of disasters. They impact everyone, including those who are supposed to assist immediately. Thus, the mental health burden in post-disaster setting is huge. In fact, a community may require all available assistance, both professional and non-professional, after an emergency. The effects of a disaster on a population may be more severe if that group is more vulnerable (Yanay and Benjamin 2005). A comprehensive disaster mental health response program must include routinely used practices like psychological first aid, crisis intervention and counselling, critical incident stress debriefing, mental health media communications, family care, resilience and community capacity building, and many other support activities (Bauwens and Naturale 2017). A number of disciplines, including social work, public health, psychiatry, and psychology, are essential in providing these mental health interventions in post-disaster environments.

10.4 Eco-Social Work—A New Approach to Social Work

As a profession, social work seeks to improve or restore the functioning of people, families, groups, and communities while also fostering social conditions that support that objective (Morales and Sheafor 1995). According to Newburn (1993), there are six social work roles that have proven beneficial in emergency situations. These include offering guidance and information, practical assistance (such as housing or income), and further therapeutic forms of involvement, like mitigation with other agencies, reconstruction, validation, and facilitation (Yanay and Benjamin 2005). The field of social work views environmental problems, including disasters, as crises in the relationship between humans and nature. Thus, it is clear that these crises cannot be resolved by means of technology and biology alone, but need changes at the societal level as well (Närhi 2004). Eco-social work seeks to fill this gap. By focusing on the well-being of individuals, the eco-social paradigm in social work aims to broaden the profession's focus to include the well-being of the entire world and all of its species. The natural environment is essential to the wellness of all people because humans are a part of nature. The study of eco-social work looks at how social workers might support human and environmental well-being and connectedness (Social Work Graduate 2023). Eco-social work is applicable at all levels of practice (e.g., micro, mezzo, macro) (Rambaree et al. 2019). In post-disaster contexts, the eco-social approach offers one of the most inclusive frameworks. According to the eco-social perspective, persons are a part of and reliant on culturally created surroundings, which comprise physical configurations as well as a network of interpersonal connections and societal structures. According to Bateson's concept of an ecology of mind, cognition arises through human interaction with the environment and through interpersonal communication within a social system (Bateson, 2000). Ecological niches for humans are essentially social, with socially constructed relationships and circumstances serving as the fundamental matrix of development from birth, and cultural, with shared meanings, values, and practices influencing experience and thought throughout life (Gómez-Carrillo and Kirmayer 2023). Various social work traditions and discourses that integrate social and ecological aspects are referred to as the eco-social approach in social work. The terms eco-social, ecological, green, and environmental social work have been used interchangeably in a variety of international social work situations in recent years. The two major theoretical traditions within the eco-social framework outlined are (1) the systems theoretical and (2) ecocritical approaches. The ecosystems approach, also known as the systems theoretical perspective, emphasizes the importance of the social environment as a fundamental foundation for human development and well-being. It also stresses the use of holistic and systemic thinking in social work interventions. The ecological problem of contemporary civilization and the ecological movement, which has been gathering momentum worldwide during the 1970s and 1980s, are two examples of the ecocritical approach in social work. This approach challenges the way of life and model of Western society as a whole, which is predicated on the exploitation of people and natural resources for financial gain and strives for constant economic expansion. The conventional "social issue" of social policy has evolved into an "eco-social issue" in tandem with ecological consciousness (Närhi and Matthies 2018).

10.5 Mental Health Support Through Eco-Social Work

We live on a planet beset by our own exponential growth and arrogation of resources to pursue often short-sighted goals. As a result, we face the existential threat of climate change, which will result in massive increases in the scale and frequency of environmental catastrophes and forced migration. Globalization has created new fault lines and fractures, and media and networking have become not only bridges between communities but amplifiers of disinformation and distrust. Attacks on minorities, racism and discrimination, the erosion of refugee rights, and growing economic inequality are among the challenges we face and among the most powerful determinants of global mental health (Kirmayer 2019). Eco-social work emphasizes the person-in-environment perspective, recognizing that the challenges faced by individuals often arise from the broader social systems and environmental contexts in which they are situated. Accordingly, eco-social work seeks to promote well-being through improvements in both the immediate and extended environments surrounding individuals. This approach underscores the importance of integrating environmental considerations when addressing personal and social problems. In post-disaster contexts, individuals who experience significant losses—such as bereavement or the destruction of physical assets—frequently struggle with disrupted perceptions and coping mechanisms. From an eco-social perspective, the environment should not be viewed solely as a passive victim of disaster; rather, it should be understood as a dynamic set of factors that may contribute to or exacerbate individual difficulties. In particular, when considering mental health, problems are often precipitated by environmental stressors. Thus, the central aim of this approach is to mitigate the adverse effects of negative environmental influences and, in doing so, foster psychological resilience and promote overall mental well-being.

In post-disaster settings, eco-social workers occupy a vital position in facilitating holistic recovery processes that address the complex interplay between individual well-being, social structures, and environmental systems. Rooted in the person-in-environment (PIE) perspective, their practice acknowledges that the psychosocial challenges experienced by disaster-affected populations—such as trauma, grief, displacement, and loss of livelihoods—are not merely individual phenomena but are deeply embedded in the disrupted ecological and social contexts. Drawing on systems theory and resilience theory, eco-social workers view recovery as a multilayered process that requires interventions at the micro, mezzo, and macro levels. At the micro level, they provide psychosocial support to individuals and families coping with bereavement, anxiety, and the disorientation caused by the disaster. At the mezzo level, they work with communities to restore social cohesion, collective efficacy, and local support networks, recognizing the protective role of community solidarity in mental health recovery. At the macro level, eco-social workers advocate for policies that integrate ecological sustainability into disaster response and reconstruction efforts, ensuring that rebuilding processes do not reproduce vulnerabilities or exacerbate environmental degradation.

Furthermore, eco-social work practice challenges traditional disaster paradigms that frame the environment solely as a passive victim of catastrophe. Instead, it highlights the reciprocal relationship between human well-being and ecological health, emphasizing that environmental degradation can precipitate or intensify psychosocial distress. In this sense, eco-social workers not only address immediate humanitarian needs but also promote long-term ecological justice by fostering sustainable practices and advocating for climate-adaptive policies. Their role, therefore, extends beyond crisis intervention to include community education on environmental stewardship, capacity-building for disaster preparedness, and collaboration with interdisciplinary teams to integrate ecological perspectives into mental health care and social policy. By mitigating the negative effects of ecological stressors and reinforcing adaptive capacities at multiple levels, eco-social workers contribute to a more comprehensive model of disaster recovery—one that prioritizes both psychological resilience and ecological sustainability.

10.6 Legislative Integration of Eco-Social Work

To institutionalize eco-social work through legislation, several reforms are required. The following recommendations draw on existing scholarly work and practice models.

10.6.1 Statutory Recognition of the Person-in-Environment Principle

Legislation should explicitly mandate that social welfare and mental health laws incorporate environmental determinants of health and well-being. For example, mandates could require assessments of environmental stressors—such as pollution, disaster vulnerability, and climate change—in social work case assessments and care plans. The person-in-environment perspective is foundational to eco-social theory, which holds that individual problems cannot be separated from broader ecological and social contexts.

10.6.2 Disaster Management Laws to Embed Eco-Social Roles

Disaster response and recovery legislation should formally incorporate eco-social workers into emergency management structures. These roles should include responsibility for both psychosocial support and advocacy for sustainable environmental recovery (e.g. rebuilding in ways that reduce future risk, restoring natural systems). This aligns with ecological social work frameworks that emphasize holistic recovery—both human and environmental.

10.6.3 Interagency and Cross-Sector Collaboration

Laws should promote structural coordination among ministries or departments of social welfare, environment, health, disaster management, and planning. Legal instruments—statutes, regulations, policies—should require joint planning, shared objectives, and coordinated implementation. This mirrors calls in the literature for interdisciplinary approaches and structural social work to engage with environmental sustainability.

10.6.4 Legal Mandates for Training, Accreditation, and Curriculum Reform

Professional regulation and social work licensure laws should require eco-social work competencies, including knowledge of environmental justice, ecological systems, climate change, and disaster resilience. Academic accreditation standards should mandate the inclusion of eco-social content in social work curricula. Such training ensures a workforce capable of implementing holistic, environment-aware social services.

10.6.5 Funding and Incentives for Eco-Social Projects

Legislation should establish funding mechanisms and fiscal incentives for interventions that simultaneously address environmental restoration, disaster resilience, and social support. This could include grants for community rebuilding, green infrastructure, sustainable housing, ecosystem protection, and climate adaptation programs led by social work agencies. This is consistent with theoretical and practice literature showing that sustainability transitions in communities require financial and policy support.

10.6.6 Incorporation of Human Rights and Environmental Justice Norms

Laws should embed the idea that environmental health is a component of human rights and that environmental justice is central to social justice. Legal provisions should protect vulnerable populations from disproportionate environmental harms (such as pollution, disasters, and climate change) and ensure participatory decision-making.

10.7 Conclusion

Mental health and well-being in post-disaster contexts cannot be fully understood or addressed without recognizing the dynamic interplay between individuals, their social systems, and the ecological environments they inhabit. The eco-social work approach offers a holistic framework that situates psychosocial recovery within the broader context of environmental sustainability and community resilience. By integrating the person-in-environment perspective with principles of ecological justice, eco-social workers are uniquely positioned to support trauma recovery, rebuild social cohesion, and advocate for sustainable reconstruction practices that reduce future vulnerabilities. This approach not only addresses immediate psychological and social needs but also fosters long-term resilience by embedding recovery efforts within environmentally responsible and socially just frameworks.

As disasters increasingly intersect with the global climate crisis, the relevance of eco-social work becomes even more urgent. Embedding eco-social principles into legislation, policy, and professional practice can institutionalize this integrative approach, ensuring that post-disaster interventions move beyond short-term relief to support transformative, sustainable recovery. Ultimately, eco-social work underscores that human well-being and ecological health are inextricably linked, and that fostering resilience in disaster-affected populations requires addressing both dimensions simultaneously. By doing so, it offers a pathway toward recovery that is not only holistic but also ethically and environmentally sustainable.

References

Bateson G (2000) Steps to an ecology of mind: Collected essays in anthropology, psychiatry, evolution, and epistemology.University of Chicago press

Bauwens J, Naturale A (2017) The role of social work in the aftermath of disasters and traumatic events. Clin Soc Work J 45(2): 99–101. https://doi.org/10.1007/s10615-017-0623-8

Goldmann E, Galea S (2014) Mental health consequences of disasters. Annu Rev Public Health 35:169–183. https://doi.org/10.1146/annurev-publhealth-032013-182435

Gómez-Carrillo A, Kirmayer LJ (2023) A cultural-ecosocial systems view for psychiatry. Front Psych 14:1031390. https://doi.org/10.3389/fpsyt.2023.1031390

Kirmayer LJ (2019) Toward an ecosocial psychiatry. World Soc Psychiatry 1(1):30–32. https://doi.org/10.4103/WSP.WSP_9_19

Morales AT, Sheafor BW (1995) Social Work: A Profession of Many Faces (7th Ed), London: Allyn andBaconhttps://www.undrr.org/terminology/disaster. Accessed 7 Aug 2025

Newburn T (1993) Disaster and after: social work in the aftermath of disaster. London: Jessica Kingsley

Närhi K (2004) The eco-social approach in social work and the challenges to the expertise of social work. Doctoral dissertation, University of Jyväskylä. https://urn.fi/URN:ISBN:951-39-1834-3

Närhi K, Matthies A-L (2018) The ecosocial approach in social work as a framework for structural social work. Int Soc Work 61(4):490–502. https://doi.org/10.1177/0020872816644663

National Institute of Disaster Management (2014) Understanding disasters. Ministry of Home Affairs, Government of India. https://nidm.gov.in/PDF/Disaster_about.pdf

Public EM-DAT Platform. Retrieved August 13, 2025 from https://public.emdat.be/

Rambaree K, Powers MCF, Smith RJ (2019) Ecosocial work and social change in community practice. J Community Pract 27(3–4):205–212. https://doi.org/10.1080/10705422.2019.1660516

Social Work Graduate (2023) Eco social work. Social Work Graduate, 27 Oct 2023. https://www.thesocialworkgraduate.com/post/eco-social-work

The Disaster Management Act, 2005, c. 1(Section 2(d))

UNDRR. (2017). The Sendai framework terminology on disaster risk reduction. "Disaster". https://www.undrr.org/terminology/disaster. Accessed 7 Aug 2025

Yanay U, Benjamin S (2005) The role of social workers in disasters: the Jerusalem experience. Int Soc Work 48(3):263–276. https://doi.org/10.1177/0020872805051704

Chapter 11
Responsible Consumption, Management and Sustainability in Fine-Dining Restaurants

Sonny Jose, Kristina Grumadaite, Angelo P. Mathew, and Sebin Thomas Babu

Abstract The fine-dining restaurant industry, considering its level of sophistication and the exclusivity of the clients it serves, has been called to account for within the framework of UNSDG Goal 12. The restaurant industry has been increasingly scrutinised given the fact that food waste alone contributes to 8% of the greenhouse gases (GHGs) and that its packaging constitutes almost 50% of the landfills. The present article looks at the problem and critiques the approach from an eco-social work perspective and applies the complex-systems theory to examine the various strategies adopted by fine-dining restaurants worldwide. A logical comparison between the good practices of the resource-rich, affluent Global North and the resource-strapped, yet rich in community engagement approaches of the Global South presents opportunities for incorporating green practices across the industry.

Keywords Eco-social work · Complex systems · Community-based participation · Green practices · Responsible consumption · Supply chain impacts · Sustainability · Sustainable development · UNSDGs

---

S. Jose
International Relations and Mobility | Loyola College of Social Sciences, Sreekariyam, Thiruvananthapuram, Kerala, India
e-mail: sonny.jose@loyolacollegekerala.edu.in

K. Grumadaite · S. T. Babu
St. Ignatius of Loyola College, Kaunas, Lithuania
e-mail: Kristina.grumadaite@ilk.lt; sebin.thomas.babu@ilk.lt

A. P. Mathew (✉)
Human Resource Management, Loyola College of Social Sciences, Thiruvananthapuram, Kerala, India
e-mail: angelo.mathew@loyolacollegekerala.edu.in

J. K. Joseph, A. P. Pradeepkumar (eds.), *Eco-Social Work for Disaster Risk Reduction, Disaster Resilience and Green Growth*, https://doi.org/10.1007/978-981-95-4885-9_11

11.1 Introduction

The restaurant sector has been increasingly scrutinized within the framework of the UN 2030 Agenda and the Sustainable Development Goals (SDGs), particularly Goal 12: Responsible Consumption and Production. The industry contributes substantially to climate change and biodiversity loss; for instance, global food waste alone accounts for approximately 8% of greenhouse gas emissions (Beck 2022). Waste from the hospitality sector, including food and packaging, constitutes nearly 50% of landfill streams in many countries. Recent analyses caution that without rapid innovation, the sector will fail to meet carbon reduction targets, given that the *World Sustainable Hospitality Alliance* calls for a 66% reduction in the industry's carbon footprint by 2030 (World Sustainable Hospitality Alliance, 2023). Simultaneously, regulatory and consumer pressures are driving change, with measures such as bans on single-use plastics, adoption of energy-efficient technologies, and commitments to halving food waste in line with *SDG 12.3* (Rosen 2014). In response, restaurants, ranging from farm-to-table enterprises to fine dining establishments, are increasingly embedding sustainability across their operations and supply chains.

A perspective inspired by *eco-social work* augments the *World Sustainable Hospitality Alliance* agenda by emphasizing human well-being and environmental health as interdependent, rather than mutually exclusive (The Social Work Graduate 2023). Thus, in practice the food industry needs to reform dining culture, incorporating ecological outcomes alongside social attitudes, community impacts, and equity considerations. A *complex-systems approach* conceptualises restaurants as nodes in a wider food-energy-waste network that encompasses farms, transport, kitchens, and households. Consequently, sustainable practices must extend across a host of domains: sourcing local produce, engaging in regenerative farming, designing low-waste menus, investing in staff training, educating consumers, and engaging communities through food donation initiatives. These principles align with eco-social education, which asserts that social justice cannot be achieved without ecological justice (Rambaree 2020). By adopting this holistic framework, practitioners and food industry actors may be able to design interventions that simultaneously address environmental sustainability and social equity within the fine-dining sector (Rambaree 2023).

11.2 Environmental and Social Impacts of Restaurant Operations

Restaurants are resource-intensive businesses. Beyond waste quantities generated, restaurants have a tremendous social impact; employment conditions, community health (e.g., diets), and food security are affected by how restaurants operate. They impact *supply chain* by buying large quantities of water-intensive or emission-intensive foods such as meat, dairy, farmed shrimp, out-of-season produce, etc., by transporting ingredients long distances, and by consuming energy in kitchens and dining rooms. On the output side, they generate large volumes of solid waste—plate leftovers, spoiled ingredients, packaging and wastewater. For example, a case study of a coastal Italian restaurant found *per meal* waste averages of 0.72 kg, of which roughly 28% was raw food waste (unused food scraps), while glass, paper, cardboard and plastic each accounted for around 17–22% of that waste (Tatàno et al. 2017). The study noted that waste intensity correlates with menu and portion practices; in other words, limiting menu items, controlling portion sizes, and using high-quality local ingredients can *significantly reduce food waste per meal*. Similarly, surveys in Southern India have highlighted that most restaurant waste comes from *preparation losses, spoilage and plate leftovers, lack of storage facilities* and *absence of waste tracking practice*, and that strategies like composting and surplus-food donation are generally well-received by staff, but hampered by lack of awareness or infrastructure—storage and waste-trackers (Sabarirajan et al. 2024).

An eco-social perspective poses crucial questions: Who benefits or suffers when a restaurant goes green? For instance, a switch to an entirely plant-based menu might boost public health, but in turn, also affect farmers or workers in the meat industry. A circular food-sharing program reduces landfill waste, but raises questions of food equity, besides quality. As one scholar notes, an eco-social work perspective requires expanding our notion of "well-being" to include environmental quality and even *justice for other species* on the planet (The Social Work Graduate 2023). In the broader eco-social justice perspective, this emphasises the factoring in of the interconnectedness of human well-being and environmental health; in the restaurant context, in a similar fashion, this translates to practices that tie environmental goals to social ones—for example, hiring local villagers, paying fair prices to traditional farmers, raising awareness amongst stakeholders, especially consumers and restaurant employees, using excess food produced to feed the needy, etc.

11.3 Responsible Consumption and Sustainability Practices in the Restaurant Industry: Case Studies

In the present article the authors study the various cases taken from the Global North and Global South, so as to keep a balance in understanding the perspective and practices. These two hemispheres differ significantly in economic development, infrastructure, cultural values, regulatory frameworks, and resource availability, all of which influence how restaurants approach sustainability (Hajmohammad et al. 2013). For example, fine-dining restaurants in the Global North often have greater access to resources and advanced technologies, e.g., composting equipment and renewable energy, whereas those in the Global South face infrastructure constraints, yet possess deep-rooted traditional knowledge and remarkable models of community engagement (Leal Filho et al. 2019). Studying both regions enables a comprehensive understanding of how context-specific factors affect sustainability efforts and responsible consumption. Further, analysing practices across both Global North and South provides the opportunity to capture the *upstream and downstream dynamics* that span geographies and affect ecological footprints worldwide. This approach aligns with complex-systems theory emphasising cross-scale interactions (Meadows 2008). The case studies appended to this segment highlight certain good practices that may be emulated by sensible fine-dining restaurants in the industry:

11.3.1 The Global North

- *London's Silo (UK)*: Widely recognized as a pioneer of zero-waste fine dining, Silo has embedded sustainability into every aspect of its business. At London's Silo, a leading zero-waste restaurant, Chef Doug McMaster offers only one set menu (touring dishes) each evening to simplify ordering, and all the ingredients required arrive in reusable containers (any disposable packaging is refused) as agreed with the farmer suppliers. The interior décor itself tells the story: tables and chairs are crafted from repurposed wood and cardboard, lighting from wine bottles, and even the plates are made of shredded plastic. The restaurant operates without a trash can; in effect, nothing is thrown "away"—residual oil is recycled into soaps or fuelling, and organic scraps become pig-feed or compost (Smithsonian Magazine 2018, 2023). Silo's very radical approach earned it a Michelin Green Star and has been a key influence for many UK restaurants (Michelin Guide 2024a). This approach eliminates the risk of unused ingredients, ensuring that all food items are utilized efficiently. Silo's design goes all out to emphasise sustainability, with plates made from recycled plastic bags and tables from packaging waste, composting all food scraps and working directly with farmers to source ingredients in reusable containers.
- *Frea (Berlin, Germany)*: As Germany's first zero-waste restaurant, Frea operates on a farm-to-table principle. It grows much of its produce in a kitchen garden and brings in ingredients via electric bicycles. Frea's hallmark is *Gersi*—its in-house developed on-site composter; all inedible scraps are fed into this machine, which churns them into usable compost within 24 h (Sagmeister 2019). By quickly cycling organics back into soil, Frea stays almost entirely waste-free. The menu at Frea is entirely plant-based, not only for ethical purposes, but also follows the logic that plants yield higher edible volumes. Besides, its deep philosophy for sustainability in the restaurant's motto is even in the table settings and signage being done out of recyclable or biodegradable materials.
- *Blue Hill, Stone Barns (USA)*: Chef Dan Barber's farm-connect themed restaurant near New York City has long been a model of regenerative gastronomy. It sources from the adjacent Stone Barns Center for Food & Agriculture, supporting over 60 local farms and breeders. Beyond conventional waste bins, Blue Hill sorts *everything*: kitchen scraps go to compost piles; animal feed is produced by

fermenting food waste; recyclable materials are collected separately (Gaskins 2023). Remarkably, the farm pigs are fed only leftover grains and vegetable trimmings, closing nutrient loops on-site. Besides, the restaurant also preserves seasonal surpluses by means of curing, fermenting and freezing. In 2020 alone, 30% of its winter menu was harvested in advance to store and use over the winter months (Gaskins 2023). This strategy supports farm resilience and demonstrates that fine dining can be deeply integrated with sustainable agriculture. Blue Hill's commitment to sustainability and ongoing research into climate-smart diets is reflected in its Michelin Green Star.

- *"Restaurant X" (Slovenia)*: A 2025 case study by Zhang et al. (2025) of a Michelin Green Star restaurant (anonymized as "Restaurant X") illustrates extreme self-sufficiency (Michelin Guide 2024b). This establishment owns a duck farm and vegetable gardens, producing many proteins and herbs on-site. It publishes seasonal menus based on what's available and practices rigorous waste reduction, e.g., composting and recycling everything possible. Interviews with the chef-owner reveal that his *lifestyle*, growing food at home, cycling to work, etc., shapes the restaurant's culture of sustainability. The study notes that "Restaurant X" had achieved 80% self-sufficiency: most ingredients came from its own operations or trusted local partners, and it cultivated a community network of farmers for the rest. Its experience demonstrates how that even high-end cuisine can align with Goals 12.2–12.5 of Agenda 2030 by efficient resource use and substantive waste reduction.
- *Farmers' Circle, Vilnius (Lithuania)*: Danish food entrepreneur Niels Peter Pretzmann established the "Baltic Food Republic" in 2015, which operates "*Senatorių Pasažas*" ("Senator's Passage"), once a Palace complete with a monk's brewery and stables, now a vibrant food hall in the Old Town Vilnius, with major farm-to-table restaurants "14Horses," and "Nineteen18," various bars and wineries, besides numerous sustainable organic vegetables and herbs food outlets produced by local sustainable farmers (Olsen 2024).
  The "*14Horses*" qualified for the prestigious Michelin Bib Gourmand in 2024 and 2025, for offering good quality food at a moderate price, while "*Nineteen18*" received its first Michelin Star for high quality cooking in 2024, and retained it in 2025 too. Pretzmann established these two restaurants, with the central idea of incorporating more organic food into the menu. However, he experienced limited options in sourcing organic raw materials from local Lithuanian farmers regularly, besides finding the produce unreliable in quality. To offset this shortage, Pretzmann bought a 600-hectare farm in 2018 in the village of Radiškis, Ukmergės district, Lithuania, where he established the "*Farmers Circle*." The "*Farmers Circle*" evolved as a regenerative farm (JSC), based on the "*soil food web*" approach that visualised farm soil as a community of diverse organisms—primary decomposers along with main predators (Harkes et al. 2019)—together providing a variety of ecological functions allowing soil regeneration (Lin et al. 2025), protecting the soil against diseases and pests, and also inhibiting the growth of weeds. To maintain the soil food web approach, "*Farmers Circle*" implements a minimum tillage (8 cm soil cultivation and without flipping over

the soil tilled), does crop rotation, by interchanging cover crops, and cultivating grassland and young wheat as a source for whole seed foods for the cows (Olsen 2024). At the time of writing the article, the "*Farmers Circle*" employed 500 Black Angus cattle, grazing over 570 hectares of pasture land, with their manure positively influencing soil quality and biodiversity, promoting soil micro-life revitalization. The farm also housed 1500 hens, which walk freely, boasts 20 hectares of veggies and has its own slaughterhouse, farm laboratory, and shopping places of organic products to promote self-reliance.

The "*Red Brick*" *restaurant* alongside a 15-room "Sleepy Horse" stable-turned-guesthouse established within the area of "Farmers Circle," offers seasonal local dishes, besides the Baltic and Nordic cuisines. It received the "Michelin Green Star" for sustainability practices 2024 and 2025, and an official silver organic certificate acknowledging that 60–90% of the products served in the restaurant are organic (Žvinakytė 2023). Regenerative farming together with "Red Brick" restaurant employs ecological, economic and social principles, and this practice, according to Pretzmann, not only ensures soil health and soil fertility and decreases gas emission but also creates a closer connection with customers, who are involved at the very beginning of the supply chain through various practices, especially through restaurant and shops.

11.3.2 The Global South

- *Locavore NXT, Bali (Indonesia)*: Opened in late 2023 in Ubud, Locavore NXT brands itself as a "localised rebellion" against wasteful practices. From day one it integrated sustainability: the property includes rooftop herb gardens, beehives, and a dedicated fermentation kitchen (Manalac 2025). Its on-site *Circular Waste Centre* (CWC) with shredders, digesters and composters capable of segregating and processing all waste streams was so efficient that its very first year, the CWC diverted an astonishing 98.4% of waste from landfill. Plastic, paper and metal are compacted for recycling; food scraps become vermicompost or animal feed. Importantly, the CWC is being expanded to serve the wider community, embodying a social dimension to its mission towards sustainability (The SRA, 2024a). The menu is 75% plant-based and uses only local, heritage ingredients, explicitly excluding beef, dairy, refined sugars or wheat (Manalac 2025). The Locavore team also recruits its staff from the local community with a hiring target of at least 20%. It also organizes community clean-ups and education on waste. In sum, Locavore NXT exemplifies holistic, community-linked sustainability in a hospitality setting.
- *Haoma, Bangkok (Thailand)*: The winner of Asia's 50 Best Sustainable Restaurant Award 2024, Haoma embeds ecology into its DNA and operates as an "urban farm oasis," where diners eat amid greenery. Haoma's philosophy of

"grow to give back" finds the restaurant sourcing most ingredients from its own organic farm, on-site crops, 500+ chickens, cows and goats, besides using harvested rainwater for cooking and irrigation (The World's 50 Best Restaurants 2023). Its rooftop farm and nearby organic fields produce the bulk of its vegetables, herbs and fruits; poultry and livestock raised on-site provide eggs and dairy. Post-renovation, Haoma also introduced water-saving systems: a rainwater harvesting network that captures 150,000 L/year for reuse in cooking, fish ponds and gardens, requiring less than 3–5% of the total water requirement from external sources. The menu is almost 75% vegan/vegetarian, and the team tracks almost every ingredient. The study by Paredes (2024), highlights Haoma's pursuit of zero waste: it keeps its kitchen waste below 5%, turning peelings into fish food or compost, and recycling cooking oil and containers via partner projects. It also supported local workers by feeding migrant communities during COVID. Importantly, Haoma's founder espouses that even small acts such as "planting one tree per diner" add up. This case shows an Asian fine-dining venue taking a farm-centric, circular approach that influences both guests and the broader food system.
- *The Aburi Sushi, Bali (Indonesia)*: This Japanese-style restaurant, part of Bali's culinary tourism circuit, offers another angle on waste prevention. A qualitative case study on "The Aburi" indicates that the team uses meticulous inventory management and portion control (e.g., pre-scaling fish cuts) to avoid over-ordering (Subadra 2024). Its menu planning and staff training are built around minimizing leftovers. Notably, "The Aburi" donates surplus meals daily and educates customers on waste reduction during service. Subadra (2024) concludes that these measures, combined with staff engagement and clear documentation, significantly reduce the restaurant's *food loss and waste*, with benefits for community well-being and brand value. This is the best example of a good practice that illustrates how detailed, data-driven management in an Asian dining context can address sustainability in practice.
- Another multi-country review notes that much waste in Asian restaurants still occurs in the kitchen due to trimmings and spoiled produce and in dining rooms due to plate waste. Hence, the best way to reduce the same was to engage in targeted interventions to mitigate waste in both kitchen operations and customer dining practices (Filimonau et al. 2020).

Thus, across Asia, there is a diversity of approaches: some upscale venues created urban herb gardens or close water loops (as Haoma did), while NGOs and startups promoted app-based food-sharing and waste tracking. The three cases, Bali Locavore, Bangkok Haoma, and Bali Aburi, all highlight the *cutting edge* of these trends, where high-end dining aligns directly with community engagement beyond ecological goals.

11.4 Strategies for Responsible Consumption and Waste Management

Given these challenges, many restaurants have attempted innovating across the *consumption–management–recycling chain*. Listed below are some of the key strategies and practices:

- *Local, seasonal sourcing*. The practice of reducing food miles by sourcing from nearby farms shrinks carbon footprints and often yields fresher produce. Menus built around regional ingredients also naturally change with seasons, lowering reliance on imported or exotic foods. As a case in Lithuania noted, working with *small local farms* and even developing heritage crop varieties can align economic incentives with ecological goals (Michelin Guide 2024a). Many green restaurants in Spain try to *minimize animal products*, incorporating more plant-centric menus to the tune of 75–100%, thereby both cutting emissions and reducing waste (Espluga-Trenc et al. 2021).
- *Portion and menu design*. Fixed, *prix-fixe* menus (or smaller plates) can prevent surplus production. For instance, London's Silo, a leading zero-waste restaurant, offers only one set menu each evening, eliminating the risk that odd ingredients go unused (Matchar 2020). Elsewhere, chefs design dishes that use whole ingredients ("root-to-stem," "nose-to-tail" cooking) so that nothing is discarded (Food and Wine 2019). Similarly, shorter menus also help kitchens plan precisely; combining this with reservation and ordering apps, so that kitchens estimate how many diners will come, is becoming more common.
- *Donation and composting*. In many countries, laws now encourage or mandate the donation of edible surplus to charities. Restaurants often collaborate with food banks or community groups to share leftovers with the needy, thus preventing food wastage. Equally important is *on-site composting or anaerobic digestion*: dozens of leading kitchens install composters or fermenters. Frea, a Berlin café built an in-house composter machine ("*Gersi*") that transforms its peels and trimmings into compost within a day (Matchar 2020). In tropical Bali, another restaurant implemented a "circular waste centre" with shredders and hot composters that diverted over 98% of all waste away from landfill in its first year (The SRA 2024). These systems often send finished compost to local farms or gardens, closing the loop. Such innovative waste reduction strategies in the hospitality industry emphasises the importance of sustainable practices in reducing landfill contributions.
- *Recycling and upcycling*. Beyond organics, fine-dining venues are finding creative uses for non-food waste. Recyclable materials are segregated and sent to appropriate facilities, while *upcycling* can add aesthetic or functional value. For example, London's Silo even made its plates from recycled plastic bags and furniture from salvaged wood and wine bottles (Food and Wine 2019). Eataly and Balthazar in New York City partner with Grease Lightning, a company that collects used cooking oil from various restaurants, recycles, and converts it into

biodiesel (Restaurant Business Online 2018; Bowen 2012). In general, the practice of consciously eliminating single-use items, e.g., switching to washable cutlery and tableware, is a straightforward step. France's legislative move to ban disposable plates, cups, and tableware for dine-in customers, mandating the use of reusable, washable alternatives in restaurants is a good practice that could be emulated (Chrisafis 2022). The Clean Water Fund's *ReThink Disposable* program has demonstrated that transitioning to reusable foodware has eliminated over 2.8 million disposable items annually, reducing waste by 14.8 tons and an average saving of $3000 per year for businesses (EPA 2025).

- *Energy and water efficiency*. Many restaurants upgrade electrical equipment, LED lights, efficient ovens/HVAC, solar panels, etc., and install smart controls. Water reuse, filtering and reusing greywater for irrigation or non-potable uses, is also an emerging good practice. Haoma in Bangkok, for instance, captures rainwater almost to the tune of 150,000 L each year for the purpose of cooking and farming, thereby reducing municipal draw by 95% (The World's 50 Best Restaurants 2023). Closed-loop designs, rain gardens and even "green roofs" (with food-growing beds) are being experimented on so as to cut heating and cooling loads.

Taken together, these good practices illustrate a systems view: changes at procurement (sourcing locally and seasonally), impact production (menu design) and waste flows (composting and donating) simultaneously, requiring coordination among chefs, managers, and sometimes even diners (van den Berg et al. 2023). Crucially, the success and longevity of such practices depend on the *team and culture* (Jaganjac and Høgevold 2024). Studies show that when restaurant owners and staff embrace sustainability as a core value, innovative practices tend to endure (Testa et al. 2020). Training and awareness programs, often led by "green champions" within staff, are common interventions in eco-social projects. A research project in India demonstrated how technological interventions, combined with staff training and awareness initiatives, have led to significant improvements in food waste management practices, including improved composting and food-safety practices in hotels (Singh et al. 2024).

11.5 Eco-Social Work: Concept and Perspective

Eco-social work is an emergent, interdisciplinary paradigm within social work that integrates ecological awareness and environmental justice into the framework of social well-being. As defined by Boetto (2017), "eco-social work refers to an integrated approach that addresses the interdependence between social systems and ecological systems, aiming to promote environmental sustainability, social equity, and human well-being." Eco-social work is positioned differently from allied concepts—*green social work*, which stresses macro-level green development and justice (Dominelli 2013), and *environmental social work* that focuses on environmental protection and justice (Ramsay and Boddy 2017).

Eco-social work integrates indigenous ecological knowledge and cultural diversity into interventions, emphasizing *harmony between humans, nature* and *welfare integration* (Boetto 2017; Matthies et al. 2019). It positions environmental sustainability as a matter inseparable from human welfare, advocating for systemic interventions that address both ecological degradation and social inequalities simultaneously (Dominelli 2012; Boetto 2017). The core premise is that the degradation of natural ecosystems disproportionately affects the most vulnerable populations, and hence, social work must commit to and actively engage with environmental issues to achieve its ethical commitments. This has implications and prospects for practice within the food industry.

Further, eco-social work builds upon critical and green social work traditions, drawing from environmental ethics, indigenous knowledge, systems theory, and sustainability science. It supports the idea that human and environmental health are co-dependent (Ramsay and Boddy 2017), and it promotes interventions that nurture regenerative relationships between people, communities, and ecosystems. The food industry is a large ecosystem consisting of farmers, suppliers, chefs, utility providers, regulators, food and material recovery systems, restaurant employees and diners, providing ample space for social work engagement.

11.5.1 Characteristics of Eco-Social Work

The essential characteristics of eco-social work are:

1. *Holistic and systems-oriented*: Eco-social work recognises and acknowledges that humans, as part of a broader ecosystem, and hence, advocates for a holistic view of social problems. Thus, it employs the complex-systems thinking to understand the interlinkages between environment, economy, and society. This encompasses the fine-dining industry with the ecosystem and the stakeholders, and offers a prospective area for social work engagement.
2. *Justice-focused*: Eco-social Work Centres on environmental justice, recognizing that marginalized communities are often disproportionately affected by environmental harms (Dominelli 2012). Hence, it aims to promote intergenerational equity, sustainability, and rights for both humans and non-human life. In the context of the food industry, it must acknowledge those who remain unfed, the exclusion of the local community, the rights of plants and animals, the over-utilisation of land, and food wastage.
3. *Context-based and culturally-sensitive*: Eco-social Work encourages context-specific solutions grounded in local ecosystems, traditions, and indigenous knowledge systems (Ife 2016). It engages communities in participatory and empowering ways, besides inclusion and promotion of traditional cuisines,

appreciating cross-cultural interactions and bringing in the dimension of universality into the context of dining.

4. *Preventive and regenerative*: Eco-social Work focuses on prevention of harm (e.g., pollution, resource depletion, exploitation, imposition of dominant cultural practices) and regenerative practices (e.g., ecological restoration, food security justice). Here too, we need to consider the prevention of pollution of the water, air and soil, besides the protection of the local flora and fauna.
5. *Interdisciplinary and collaborative*: Eco-social Work works intersectorally, bringing in the rich know-how from sectors such as public health, urban planning, agricultural and nutritional sciences, environmental science, and sustainable restaurant business.
6. *Ethical and Reflexive*: Eco-social Work calls for critical reflection on excessive consumerism, capitalist growth models, and the profession's own environmental footprint.

11.5.2 Situating Eco-Social Work Principles in the Context of the Fine-Dining Restaurant Industry

The fine-dining industry, traditionally associated with luxury and resource intensity, has come under progressively scrutinised in the context of sustainability and the *UN Sustainable Development Goals (SDGs)*, especially SDG 12: Responsible Consumption and Production. Eco-social work has strong conceptual and practical relevance in helping this industry align with environmental and social responsibility. The below-given table shows how the principles of Eco-social Work may apply in various contexts within the Fine-dining Restaurant business. Table 11.1 illustrates the practical application of eco-social work principles within the context of fine-dining operations.

Hence, it goes without saying that Eco-social Work does have significant implications for the Restaurant industry.

11.6 Implications of Eco-Social Work Practice for Hospitality

The earlier cited examples carry lessons for eco-social practitioners working with communities with a possibility of associating with the hospitality sector. To begin with, *education* and *engagement* are key: diner attitudes influence restaurant norms, and vice versa. Social workers and community organizers can help launch awareness community education campaigns (e.g., explaining portions, the value of leftovers, or the carbon impact of food choices) in partnership with local eateries. They may consider also facilitating programs such as "*pay-what-you-can*" or

Table 11.1 Eco-social work applications in fine-dining sustainability practices

Eco-social work principle	Application in fine-dining sustainability
Holistic systems thinking	Restaurants may be considered as components of the food-energy-waste networks; hence, they need to acknowledge the role of the various stakeholders, farmers, suppliers, diners, and waste handlers, and link them up. Good practice includes low-waste menu design, closed-loop composting, and zero-mile sourcing.
Social and environmental justice	It is considered ethical to support fair labour, advocate and provide for equitable food access (e.g., surplus food donation), and environmental justice (e.g., reducing emissions from food waste and transport).
Community engagement	Encourage partnerships with all stakeholders—local farmers, suppliers, NGOs, and food rescue groups; as a good practice it is important to involve diners through sustainability messaging and participatory awareness campaigns.
Place-based innovation	Promote the use of heritage crops, local varieties, seasonal crops and fruits, besides adopting traditional cooking methods, so as to reduce carbon footprints while maintaining cultural integrity.
Ethical and reflexive practice	Guide restaurants and their key stakeholders need to be proficient and transparent in self-assessing their ecological impacts (e.g., water and energy use) and social responsibilities (e.g., staff training, inclusive sourcing, environmental impact of their operations, etc.)
Regenerative approaches	Encourage practices like regenerative agriculture partnerships, biodiversity promotion (e.g., pollinator gardens), and composting to rebuild soil and ecosystems.

meal-sharing for excess food, linking restaurants to food banks or even directly to vulnerable groups. Second, is policy and advocacy; here eco-social professionals can inform policymakers about the importance of green regulations (mandatory composting programs, incentives for renewable energy in restaurants, etc.), ensuring that hospitality workers are included in training for green skills.

At the organizational level, our analysis suggests that *stewardship* and *culture change* are critical. An eco-social approach would encourage restaurant owners and staff to see themselves as stewards of both *people* and *planet*; for instance, the organisation may consider incentives, embedding sustainability metrics into performance reviews or by celebrating "green chef" roles. Complex-systems thinking implies that interventions should cut across "traditional" boundaries. For instance, training kitchen staff in recycling is important; but pairing that with community tree-planting days or nutrition education in the neighbourhood, besides amplifying its impact, takes their social responsibility to the next level. Eco-social work often emphasizes resilience and equity: thus, supporting small farmers, who supply green restaurants, ensuring that they earn a fair income while adopting sustainable practices, falls squarely under this domain.

Finally, in the context of research and monitoring, social workers applying *community-based participatory research* in restaurant settings can help assess what works. For example, social workers could co-develop waste-audit projects with restaurants, or survey the diners' willingness to adopt responsible consumption behaviours such as ordering smaller portions or accepting "ugly" produce. The case studies show that data—waste measurements, compost diversion rates, etc.—is a powerful tool to motivate change. By acting as bridges between environmental science and social work practice, eco-social professionals can help institutionalise the positive examples illustrated above and adapt them to less-resourced settings.

11.7 Complex-Systems Theory and Eco-Social Work in Fine Dining: A Critical Analysis

11.7.1 Conceptualising Complex-Systems Theory

The complex-systems theory posits that systems are composed of numerous interdependent components, whose interactions give rise to emergent properties—outcomes that cannot be understood by analysing parts in isolation (Meadows 2008; Capra and Luisi 2014). In the context of sustainability, this theory is crucial for understanding how changes in one part of a socio-ecological system—such as a restaurant's sourcing practices—can ripple across food production, energy use, cultural tensions, community well-being, and even global emissions.

11.7.2 Eco-Social Work as a Systems-Oriented Practice

Eco-social work aligns quite naturally with complex-systems thinking. It embraces interdependence between ecological and social systems, recognises the feedback loops (e.g., how waste affects health; how labour conditions affect food quality, etc.), and promotes interventions that acknowledge both environmental limits and human needs (Boetto 2017; Dominelli 2012). Applied to fine dining, eco-social work helps stakeholders move beyond the singular aim being "green" fixes, to holistic sustainability—transforming how restaurants source, prepare, serve, and dispose of food, while also addressing labour equity, encouraging sustainable farming practices-soil, water, and ecology, community engagement, and environmental justice.

11.7.3 Fine-Dining Restaurants as Complex Systems

To begin with, fine-dining operations are not just standalone units. A threadbare analysis reveals them to be situated within complex multi-scalar systems. These include:

- *Upstream systems* encompassing all the processes and entities involved in sourcing and supplying raw materials and ingredients before they reach the restaurant—farms, fisheries, distributors, suppliers, logistic providers, and intermediaries responsible for harvesting, producing, transporting, and delivering food and other resources necessary for restaurant operations, etc.
- *Internal systems* consisting of in-house operations and processes that transform inputs into final products in the form of plated food or takeaways; examples include kitchen workflows (preparation, cooking, plating), staff behaviour and training (sustainability practices, hygiene), waste management (sorting, composting, recycling), energy use, and inventory control. Internal systems directly influence efficiency, quality, and sustainability of the restaurant operations.
- *Downstream systems* are the components and activities that occur after the restaurant serves food, including customer consumption patterns, plate waste generation, food donation initiatives, recycling, composting, wastewater treatment, and the wider social and environmental impacts of dining. Downstream systems also include interactions with the community and broader food-energy-waste networks that extend beyond the restaurant premises, e.g., diners, food waste recovery, public perceptions, etc.

Considering that fine-dining restaurants possess a networked structure, they are ideal case studies for complex-systems thinking. Hence, very small changes, such as adopting zero-waste menus, can lead to non-linear, emergent effects, such as:

- Stronger ties to local farmers and customers
- Reduced landfill pressure and a cleaner environment
- Enhanced consumer awareness
- Public engagement leading to policy advocacy for sustainable food systems

To go by examples discussed earlier:

Silo (London) reconfigures every step of its value chain—using a one-set menu to limit wastage, contracting reusable supplier packaging, recycling using on-site composters, and incorporating recycled décor. These small innovations ripple across its suppliers, diners, and peer restaurants, fostering system-wide change (Matchar 2020). *Locavore NXT* (Bali), with its brand of "localised rebellion," has perfected the use of its on-site *Circular Waste Centre* (CWC) to the extent of reducing wastage by 98.4%, besides incorporating a 75% plant-based menu procured locally. More importantly, the CWC is being expanded to serve the wider community, embodying a social dimension to its mission towards sustainability. Locavore NXT also organizes community clean-ups and education on waste to engage the community to take its noble rebellion to the next level.

11.7.3.1 Silo and Locavore NXT as Models of Eco-Social Work Good Practice: An Analysis

As stated earlier, eco-social work approach integrates ecological and social justice concerns, emphasizing interconnected well-being across human and environmental systems based on the principles of *holistic, systemic thinking*; *sustainability and resilience*; *equity and justice*; and *empowerment and participation*. Let us examine how Silo and Locavore NXT embody these principles:

1. Their strategies reflect *holistic systems thinking*, addressing supply chains, kitchen operations, and waste management collectively, rather than in isolation.
2. Both promote *environmental sustainability* by drastically reducing waste, closing nutrient loops, and minimizing resource inputs.
3. The *social dimension* is addressed by fostering relationships with suppliers and the community (e.g., local sourcing, sharing surplus food, community composting).
4. They *empower* staff and patrons through transparent sustainability practices and education, encouraging broader behavioural change beyond their restaurant and business, e.g., lifestyle, diet.
5. Both of the cited restaurants embody *resilience* by designing adaptive systems that reduce dependency on linear resource flows (traditional sourcing) and mitigate environmental risks.

By embedding sustainability at multiple levels of the food system, Silo and Locavore NXT illustrate how a *complex-systems approach* can operationalise *eco-social work* principles in fine dining. Their innovations create self-reinforcing cycles of responsible consumption, resource management, and social-environmental equity, demonstrating scalable models for the hospitality industry globally.

11.7.4 Eco-Social Practices and System Leverage Points

Drawing from Meadows' (1999) framework of leverage points in systems, eco-social work targets deep leverage points, such as:

(a) *Changing mindsets and paradigms*: Eco-social practices encourage a paradigm shift: moving from extractive consumption to regenerative relationships with food, labour, and nature. Example: Haoma (Bangkok) exemplifies a mindset shift by treating dining as a stewardship act—growing its own food, using harvested rainwater, and planting one tree per guest (The World's 50 Best 2024).

(b) *Redesigning information flows*: Transparency in sourcing, waste, and social impacts (e.g., carbon menus, traceable ingredients) helps diners make informed choices and fosters accountability. Example: Blue Hill (NY) uses storytelling and real-time feedback loops between farm and kitchen to optimize resource flows and minimize waste (Michelin Guide 2024b).

(c) *Redefining system goals*: Many of these ethical restaurants move from profit-maximization to sustainability and community resilience, thereby moving toward a different level of value generation. For example, Frea (Berlin) does not just "serve food"; it aims to be "zero-waste" by composting all scraps and rejecting unsustainable suppliers, thereby redefining success in the fine-dining model (Smithsonian Magazine 2023).

11.8 Emergent Outcomes and Feedback Loops

The emergent outcomes and feedback loops are important considerations in eco-social work practice. By imbibing incrementally, the principles of eco-social work, fine-dining restaurants could evolve into nodes of innovation and resilience in larger socio-ecological systems.

Key emergent outcomes include:

1. Nutrient Cycling: Food waste turned into compost, animal feed, or biofuel loops nutrients back into the ecosystem.
2. Community Engagement: Surplus food donations, staff training in sustainability, and public education.
3. Cultural Shifts: Consumers come to expect sustainable and ethical standards, shaping broader consumption and fine-dining market behaviour.

These feedback loops create adaptive capacity: the ability to evolve and withstand external shocks, such as supply chain disruptions or environmental regulations. Perhaps another critical aspect one needs to consider is the importance of customer satisfaction. Such satisfaction may emerge from the realisation of being part of a larger mission. This enhances customer loyalty, branding and sustainability in the long run.

11.8.1 Policy and Replication Potential

Eco-social work helps scale localized innovations by influencing policy and replication. Silo, Blue Hill, Haoma and Locavore NXT are taken as cases for the present study bear witness to the transformative experiences:

Silo's zero-waste model has influenced conversations at the municipality level on circular economy practices in hospitality across London and the UK. By refusing disposable packaging and collaborating with suppliers on closed-loop delivery, Silo has demonstrated scalable ways to reduce packaging waste. Such practices are referenced by sustainability advocates and local councils in considering bans on single-use plastics and incentives for reusables in supply chains. Thus, Silo operates not just as a restaurant but also as a policy laboratory, showing how legislation could embed waste-free principles in urban food systems (Matchar 2020).

Blue Hill (New York) integrates farm-to-table sustainability with food policy advocacy. Chef Dan Barber actively engaged with U.S. policy debates on food waste, regenerative agriculture, and sustainable diets, bringing high-level visibility to the link between gastronomy and agricultural policy. Its Green Star recognition by Michelin has provided a platform to replicate its model internationally. Moreover, Blue Hill's preservation and composting practices have informed regional farm and waste management discussions in New York State, showing how fine-dining innovation can inspire municipal composting programs and farm subsidies for regenerative practices (He 2024).

Haoma's (Bangkok) philosophy of "grow to give back" has positioned it as a regional model in Asia for integrating sustainability with social equity. Its practices—rainwater harvesting, self-sufficient urban farming, and a "tree per diner" campaign—have inspired Bangkok city officials and regional NGOs to look at urban agriculture and green dining incentives. In particular, Haoma's near-zero-waste model aligns with Thailand's national roadmap on food waste reduction (SDG 12.3), positioning it as a demonstration site for replication in other urban restaurants. Its practices show policymakers that sustainability can be implemented in dense urban contexts, not just rural or suburban farms (Paredes 2024).

The success of *Locavore* NXT in Bali in waste diversion and infrastructure (Circular Waste Centre, composters, recycling, etc.) provided empirical grounding for local governments to mandate or subsidise waste management infrastructure in restaurants, especially in tourism-intensive regions. Its model designs that integrate rainwater harvesting, passive cooling, etc., influenced building regulations to require or incentivise sustainable design in new commercial premises. The use of heritage breeds, support for small farms, permaculture, and local sourcing inspired guide subsidies or certification programs promoting local/organic/heritage agriculture. *Locavore* demonstrated that sustainable gastronomy can become part of tourism policy (e.g., green tourism certification, sustainable gastronomic trails), which can create broader demand and support for such models. Its high visibility and recognition made it possible for standards or certifications (e.g., "sustainable restaurant," "zero waste," "green restaurant") to be developed or strengthened in Indonesia, inspired by benchmarks set by Locavore NXT.

Thus, one cannot underestimate the potential of fine-dining case studies to inform municipal regulations and state policy on a large scale regarding food donation, composting incentives, and packaging bans.

11.9 Challenges and System Constraints

Despite its promise, systemic change faces barriers that can counteract its advantages. Some of these are:

1. Institutional inertia: Many restaurants operate within rigid financial and cultural constraints.
2. Information silos: Lack of shared data between restaurants, suppliers, and waste handlers inhibits coordination.
3. Short feedback loops: Diners and staff may not see the long-term environmental impact of their actions.
4. Consumer Inertia: Consumers, who engage in a fine-dining experience or even stakeholders of the industry, may resist or even refuse to accept behavioural change or lifestyle modification. Those who accept need not sustain these on their own. In short, it requires consistent supportive work.

Eco-social work has the potential to offer a lens to navigate and transform these constraints by way of community engagement and putting to use participatory processes, education, and advocacy. Applying complex-systems theory reveals that eco-social work practices in fine dining are not merely add-ons—they are catalysts of systemic transformation. By targeting mindsets, feedback loops, and goals, these practices build sustainability into the very architecture of food service. This may result not only in responsible consumption and resource management but also in a more resilient, just, and regenerative ecosystem culture that will be more sustainable.

11.10 Conclusion

The restaurant industry is at a crossroads between tradition and transformation. Under the pressure of the 2030 Agenda, climate targets and changing consumer values, even fine-dining establishments are redefining their role. The examples from Europe, North America and Asia reviewed demonstrate that "green" restaurant practices are no longer niche curiosities but represent a broad movement. To succeed, restaurants must adopt holistic strategies, from farm partnerships and local hiring to waste-regeneration centres and energy upgrades, that align economic viability with environmental ethics. Importantly, a complex-systems approach coupled with an eco-social perspective reminds us that this transformation is as much about people as it is about resources: training, community integration, and social equity must accompany technical fixes.

The case studies of Silo, Blue Hill, Locavore NXT, Haoma and others show that substantial waste reduction and sustainability are achievable without compromising quality or profitability. They also show that chefs and diners can be agents of change in the wider food system. For scholars and practitioners of social work, the restaurant industry presents a unique arena: one where consumption behaviours, labour practices and ecological stewardship intersect. Future eco-social initiatives might include developing sustainable menu guidelines, advocating for fair trade sourcing, or creating "Green Table" certifications. As one commentary urges, the survival of communities (and species) depends in part on reimagining how we feed ourselves, and restaurants, at the nexus of culture and consumption, have a leading role to play.

References

Beck J (2022) This Berlin restaurant turns food waste into soil with a machine called "Gersi". Smithsonian Magazine. https://www.smithsonianmag.com/innovation/this-berlin-restaurant-turns-food-waste-into-soil-with-a-machine-called-gersi-180980438/

Boetto H (2017) A transformative eco-social model: challenging modernist assumptions in social work. Br J Soc Work 47(1):48–67. https://doi.org/10.1093/bjsw/bcw062

Bowen A (2012) Gourmet grease: cooking oil from New York's best restaurants converted to biofuel. Metro US. https://www.metro.us/gourmet-grease-cooking-oil-from-new-yorks-best-restaurants-converted-tobiofuel/

Capra F, Luisi PL (2014) The systems view of life: a unifying vision. Cambridge University Press, Cambridge

Chrisafis A (2022) Ban on single-use restaurant tableware hailed as fast-food 'revolution' in France. The Guardian. https://www.theguardian.com/world/2022/dec/28/france-ban-on-single-use-restaurant-tableware-hailed-as-fast-food-revolution

Dominelli L (2012) Green social work: from environmental crises to environmental justice. Polity Press, Cambridge

Dominelli L (2013) Environmental justice at the heart of social work practice: Greening the profession. International Journal of Social Welfare, 22(4), 431–439. https://doi.org/10.1111/ijsw.12024

Environmental Protection Agency (2025) Food service-related source reduction programs. U.S. Environmental Protection Agency Website. https://www.epa.gov/trash-free-waters/food-service-related-source-reduction-programs

Espluga-Trenc J, Calvet-Mir L, López-García D, Di Masso M, Pomar A, Tendero G (2021) Local agri-food systems as a cultural heritage strategy to recover the sustainability of local communities: insights from the Spanish case. Sustainability 13(11):6068. https://doi.org/10.3390/su13116068

Filimonau V, Zhang H, Wang L (2020) Food waste management in Shanghai full-service restaurants: a senior managers' perspective. J Clean Prod 258:120975. https://doi.org/10.1016/j.jclepro.2020.120975

Food & Wine (2019) This restaurant won't have a single trash can. https://www.foodandwine.com/news/douglas-mcmaster-silo-no-waste-restaurant

Gaskins T (2023) Eco-friendly dining destinations. Michelin Guide. https://guide.michelin.com/us/en/article/features/eco-friendly-green-star-restaurants-michelin-guide

Hajmohammad S, Vachon S, Klassen RD, Mihov M (2013) Safety or sustainability? How organizational culture influences the implementation of sustainability and health and safety initiatives. J Bus Ethics 117(3):569–589. https://doi.org/10.1007/s10551-012-1530-3

Harkes, P., Suleiman, A. K. A., van den Elsen, S., De Haan, J., Holterman, M., Kuramae, E., & Helder, J. (2019). Conventional and organic soil management as divergent drivers of resident and active fractions of major soil foodweb constituents. Scientific Reports, 9, Article 13521. https://doi.org/10.1038/s41598-019-49854-y

He M (2024) Chef Dan Barber's urgent food revolution. Michelin Guide – Sustainable Gastronomy. https://guide.michelin.com/us/en/article/sustainable-gastronomy/michelin-guide-green-star-blue-hill-at-stone-barns-new-york-sustainability

Ife J (2016) Community development in an uncertain world: vision, analysis and practice. Cambridge University Press, Cambridge

Jaganjac N, Høgevold NM (2024) The role of organizational culture and structure in implementing sustainability initiatives. Bus Ethics Environ Responsib 33(1):123–135. https://doi.org/10.1111/beer.12710

Leal Filho W, Azul AM, Brandli LL, Özuyar PG, Wall T (eds) (2019) Sustainable food systems: concepts, principles, and examples. Springer

Lin Y, Yi, Q, Gao D, Li, J, & Zhang, W. (2025) Soil micro-food-web composition determines soil fertility and crop growth. Soil Ecology Letters. https://doi.org/10.1007/s42832-024-0264-0

Manalac S (2025) A world with (almost) no waste. FoodieS. https://foodies.id/a-world-with-almost-no-waste/foodies.id+1foodies.id+1
Matchar E (2020) The rise of 'zero-waste' restaurants. Smithsonian Magazine. https://www.smithsonianmag.com/innovation/rise-zero-waste-restaurants-180974394/
Matthies, A.-L., Stamm, I., Hirvilammi, T., & Närhi, K. (2019) Ecosocial innovations and their capacity to integrate ecological, economic and social sustainability transition. Sustainability, 11(7), 2107. https://doi.org/10.3390/su11072107
Meadows DH (1999) Leverage points: places to intervene in a system. The Sustainability Institute
Meadows DH (2008) Thinking in systems: a primer. Chelsea Green Publishing
Michelin Guide (2024a) From nature to the plate: Lithuania's sustainable gastronomy. Michelin Guide. https://guide.michelin.com/en/article/sustainable-gastronomy/lithuania-from-nature-to-the-plate-red-brick-14horses-nineteen18-the-farmers-circle
Michelin Guide (2024b) Sustainable gastronomy: Blue Hill at Stone Barns. https://guide.michelin.com
Olsen, B. (2024). Where to eat and drink in Vilnius, Lithuania. National Geographic Traveller 2024, November 16. https://www.yahoo.com/news/where-eat-drink-vilnius-lithuania-090000976.html
Paredes I (2024) How Haoma in Bangkok lives its philosophy of 'grow to give back'. The World's 50 Best Restaurants. https://www.theworlds50best.com/stories/News/haoma-bangkok-sustainable-restaurant-award.html
Rambaree K (2020) Environmental social work: implications for accelerating the implementation of sustainable development in social work curricula. Int J Sustain High Educ 21:557–574
Rambaree K (2023) The responsibilities of social work for ecosocial justice. Sustainability 15(11):589. https://doi.org/10.3390/su15110589
Ramsay S, Boddy J (2017) Environmental social work: a concept analysis. Br J Soc Work 47(1):68–86. https://doi.org/10.1093/bjsw/bcw078
Restaurant Business Online (2018) Steal this idea: Restaurant turns cooking oil into soap to sell in-house. https://www.restaurantbusinessonline.com/steal-idea-restaurant-turns-cooking-oil-soap-sell-house
Rosen RJ (2014) The enormous scale of food waste in the hospitality industry. Smithsonian Magazine. https://www.smithsonianmag.com/smart-news/food-waste-hospitality-industry-180950249/
Sabarirajan A, Arunfred N, Marin VB, Sanyal S, Byloppilly R, Regin R (2024) A case study on food waste reduction strategies in Dindigul restaurants using the food supply chain. In: Singh S, Rajest SS, Hadoussa S, Obaid AJ, Regin R (eds) Data-driven intelligent business sustainability. IGI Global, pp 14–28. https://doi.org/10.4018/979-8-3693-0049-7.ch002
Sagmeister J (2019) Zero Waste: Frea Restaurant (fast) ohne Müll. Deutschlandfunk Nova. https://www.deutschlandfunknova.de/beitrag/zero-waste-frea-restaurant-fast-ohne-muell
Singh V, Archana T, Singh A, Tyagi PK (2024) Utilizing technology for food waste management in the hospitality industry hotels and restaurants. In: Singh A, Tyagi PK, Garg A (eds) Sustainable disposal methods of food wastes in hospitality operations. IGI Global, pp 341–357. https://doi.org/10.4018/979-8-3693-2181-2.ch019
Smithsonian Magazine (2018) The rise of 'zero-waste' restaurants. https://www.smithsonianmag.com/innovation/rise-zero-waste-restaurants-180974394/
Smithsonian Magazine (2023) How some restaurants are reinventing waste. https://www.smithsonianmag.com
Subadra IN (2024) Managing restaurant's food loss and waste at the Aburi Sushi Bali: a sustainable culinary tourism enterprise model in Bali. Jurnal Kajian Bali (J Bali Stud) 14(1):191–214. https://ojs.unud.ac.id/index.php/kajianbali/article/view/10722
Tatàno F, Caramiello C, Paolini T, Tripolone L (2017) Generation and collection of restaurant waste: characterization and evaluation at a case study in Italy. Waste Manag 61:423–431. https://doi.org/10.1016/j.wasman.2016.12.041

Testa F, Iraldo F, Vaccari A, Ferrari E (2020) Sustainability implementation in restaurants: a comprehensive model of drivers, barriers, and competitiveness-mediated effects on firm performance. Int J Hosp Manag 87:102510. https://doi.org/10.1016/j.ijhm.2020.102510

The Social Work Graduate (2023) Eco social work. The Social Work Graduate. https://www.thesocialworkgraduate.com/post/eco-social-work

The SRA (The Sustainable Restaurant Association). (2024) Asia's 50 Best recognises sustainability at this Balinese restaurant. The SRA. https://thesra.org/news-insights/news/asias-50-best-recognises-sustainability-atthis-balinese-restaurant/

The SRA (The Sustainable Restaurant Association) (2024a) Locavore NXT: a circular restaurant model. https://www.thesra.org

The World's 50 Best Restaurants (2023) Haoma: The Bangkok restaurant growing its own food and recycling rainwater. https://www.theworlds50best.com/stories/News/haoma-bangkok-restaurant-growing-own-food-recycling-rainwater.html

The World's 50 Best Restaurants (2024) Haoma: sustainable fine dining in Bangkok. https://www.theworlds50best.com

World Sustainable Hospitality Alliance. (2023). Global Hotel Decarbonisation Report: Pathways to carbon reduction. https://sustainablehospitalityalliance.org/resource/global-hotel-decarbonisation-report/

van den Berg L, van der Lans IA, van Trijp HCM (2023) Effectively implementing healthy and sustainable food practices in out-of-home food service locations: the perspective of the catering staff members. Appetite 187:106548. https://doi.org/10.1016/j.appet.2023.106548

Zhang Y, Li H (2025) The contribution of sustainable practices to the creation of memorable customer experience: empirical evidence from Michelin Green Star restaurants. Int J Hosp Manag 126:104110. https://doi.org/10.1016/j.ijhm.2025.104110

Žvinakytė, I (2023) Icelandic chef's 'dream' restaurant on Lithuanian farm aims for Michelin star. LRT.lt. , October 28, 2023 https://www.lrt.lt/en/news-in-english/19/2111458/icelandic-chef-s-dream-restaurant-on-lithuanian-farmaims-for-michelin-star

Chapter 12
Cross-Cutting Themes: Future Directions in Eco-Social Work and Disaster Risk Reduction

Joice K. Joseph and A. P. Pradeepkumar

Abstract This final chapter examines cross-cutting themes that shape the future directions of eco-social work and disaster risk reduction, moving beyond the conventional structural social work approach. Eco-social work, as a developing subfield, offers an innovative framework for addressing the multi-layered and intricate vulnerabilities of today's interconnected world. An eco-social lens is particularly important for responding to contemporary challenges such as the climate crisis and humanitarian emergencies. We identify ten overarching themes for future research and intervention. These are grouped into three broad areas of *environmental priorities*: eco-social work in the Anthropocene, nature-based solutions and environmental stewardship, climate change adaptation and justice, sustainability, sustainable waste management, and green disaster recovery; *social priorities*: development-induced displacement, humanitarian interventions, and responses to complex emergencies; and *educational and collaborative priorities*: advancing education and research, and strengthening multidisciplinary collaboration. Together, these themes provide a framework for advancing eco-social work practice and research in disaster contexts.

Keywords Eco-social work · Contemporary vulnerabilities · Anthropocene · Future interventions

---

J. K. Joseph (✉)
Department of MSW Disaster Management, Loyola College of Social Sciences (Autonomous), Thiruvananthapuram, Kerala, India

A. P. Pradeepkumar
Department of Geology, University of Kerala, Thiruvananthapuram, Kerala, India

J. K. Joseph, A. P. Pradeepkumar (eds.), *Eco-Social Work for Disaster Risk Reduction*, Disaster Resilience and Green Growth, https://doi.org/10.1007/978-981-95-4885-9_12

12.1 Introduction

In light of the increasingly complex challenges and sustainability issues facing the contemporary world, transdisciplinary collaboration has become essential for developing effective solutions (Hoffmann-Riem et al. 2008). To integrate social justice, community resilience, and environmental sustainability into comprehensive strategies, it is necessary to review the cross-cutting themes in eco-social work and disaster risk reduction.

12.1.1 The New Eco-Social World

Eco-social theory emphasizes how human well-being and planetary health are interconnected and considers humans as vital parts of larger ecosystems. The new paradigm calls for a redefinition of the social contract between citizens and the state and for the integration of social, economic, and ecological dimensions (Chang et al. 2025). The policy document of the International Federation of Social Workers clearly outlines the urgent need for a new eco-social world to tackle the contemporary global challenges, by focusing on the five dimensions of sustainability: people, planet, prosperity, peace, and partnership (IFSW 2022). Social workers' involvement in the eco-social domain, particularly with regard to disaster risk reduction, is multifaceted and encompasses community-focused risk reduction, resilience building, and incorporating of social and environmental justice concepts into preparedness and recovery efforts (Breen et al. 2023; Hussain et al. 2019). Global agendas like the Sendai Framework for Disaster Risk Reduction 2015–2030 and the Sustainable Development Goals (SDGs) collectively underline the need to address disaster risk and social work within sustainable development processes. Both support integrated, cross-sectoral strategies for risk management, environmental stewardship, and social justice in building community resilience. Figure 12.1 illustrates a few emerging areas that are identified as research priorities in eco-social work and disaster risk reduction.

12.2 Eco-Social Work in the Anthropocene

First proposed in 2000, the Anthropocene idea emphasises how humans are a dominating geological and environmental force, with significant consequences for accelerated mass casualties and escalating disasters (Corlett 2015; Panagiotaros et al. 2022). There is clearly an urgent need to understand the links between biophysical and social processes of the Anthropocene. Social scientist Hackmann argues that the global environmental changes are caused by humans, harm humans, and can only be solved by humans (Hackmann et al. 2014). Anthropocene is a contemporary crisis

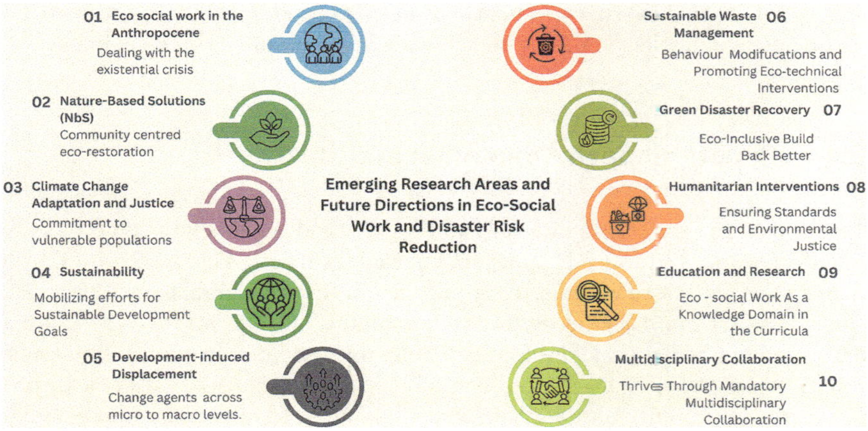

Fig. 12.1 Emerging areas of eco-social work and disaster risk reduction

which impacts all eco-social domains and is increasingly viewed as an existential crisis. In this context, the eco-social work domain must think beyond traditional neoliberal structures and plan activities such as advocating for marginalised groups facing environmental risks, fostering community resilience to climate and ecological crises, facilitating local solutions to climate change, supporting resilient livelihoods, fostering environmental literacy, practice-based research on indigenous knowledge systems, questioning overconsumption, and encouraging minimalist, mindful living.

12.3 Nature-Based Solutions (NbS)

Since nature-based solutions provide resilient, affordable, and sustainable means of tackling interrelated problems, they are essential for tackling the world's critical eco-social crises. Restoring ecosystems is essentially restoring communities; for example, a wetland restoration can reduce flood risk and promote livelihoods, or slope stabilization practices can protect soil and improve farming. Community-cantered methods and sustainable environmental management can be combined through the NbS and eco-social work integration (Dick et al. 2020). Important strategies for improving community resilience to climate impacts are provided by NbS. Numerous NbS measures are directly connected to livelihood and community health. Eco-social work can include local communities in the decision-making process and making sure the NbS solutions are locally relevant and socially inclusive (Wellmann et al. 2023). Green jobs are often created by NbS and several NbS activities have their origins in native knowledge systems. Eco-social work essentially offers the theoretical and ethical framework that guarantees NbS are not only successful but also inclusive, egalitarian, and powerful. It advocates for a people-first strategy. Environmental stewardship can be taken as a guiding philosophy being put into practice through NbS and eco-social work or can ensure its human dimensions.

12.4 Climate Change Adaptation and Justice

Climate change is impacting our day-to-day life, and vulnerable people or those with less resources and social capital are affected the most. Many professions, from science to social sciences, are actively engaged in fostering climate resilience, and eco-social work similarly needs to develop context-specific practice models that contribute to effective strategies for combating this hazard. The human dimensions of this climate crisis, such as forced migration, public health crises, and the disproportionate vulnerability of marginalized communities, have become impossible to ignore (Nyahunda et al. 2024). Climate change, traditionally characterized as a slow-onset disaster due to its gradual and snowballing impacts, is increasingly manifesting as a sudden-onset crisis, as recent extreme climate events demonstrate unforeseen and severe ecological and social disruptions like the recent Wayanad landslides in the Kerala state in India.

Promoting adaptation is one of the eco-social work's primary tasks in relation to climate change. This entails assisting people and communities in developing the resilience required to endure and bounce back from the effects of climate change. The transition can be facilitated by social workers at both the macro and mezzo levels (Sofia et al. 2024). In the Person-In-Environment (PIE) framework, the environment is often interpreted as purely social rather than encompassing the natural world, and the social work profession, through the NASW, acknowledges an obligation to address environmental protection and justice.

Climate change adaptation remains detached from social work practice when its impacts on vulnerable populations and its implications for advancing a just society are left unacknowledged and unexamined (Berger and Kelly 1993). Eco-social work practice that aims to meet the various social, psychological, and physical demands brought on by environmental changes. Innovative projects like community gardens and energy efficiency projects are all important responsibilities of eco-social workers. Additionally, eco-social workers can provide emotional support to help people and communities deal with the psychological stress and hopelessness associated with climate change, such as that experienced by climate refugees.

Climate justice demands accountability, equity, and fairness in the global climate regime, going beyond adaptation and mitigation. It is a framework that views climate change not just as an environmental or scientific issue, but as a social, ethical, and political one. Eco-social work brings essential perspectives and approaches to this challenge through its commitment to vulnerable populations, its theoretical frameworks emphasizing person-in-environment and social justice (Gray and Coates 2012). We actually needed a justice-cantered climate adaptation with bottom-up, participatory approaches.

12.5 Sustainability

Many current social work practice models do not sufficiently address environmental sustainability issues (Schmitz et al. 2012). Global issues related to environmental sustainability, such as climate instability, are intimately associated with social vulnerabilities like poverty and violence. Eco-social workers have a moral responsibility to ensure sustainability and surely humans exist in symbiotic interaction with the physical environment (Coates 2003; Cordoba and Bando 2021; Rice et al. 2022). By tackling social injustice and environmental degradation, eco-social work can advance sustainable development. The Integration of the United Nations Sustainable Development Goals (SDGs) should be a major cross-cutting theme for eco-social work, and sustainability needs to be included in the social work curriculum. The key social work skills and competencies, such as advocacy, community organizing, policy engagement, and interdisciplinary collaboration, will enable professionals to address complex social-environmental challenges of sustainability. The core competencies outlined by the Council on Social Work Education (CSWE)—such as ethical and professional behaviour, engaging with diversity, advancing human rights, and participating in policy practice—are closely aligned with sustainability. Together, they equip eco-social workers to address urgent sustainability challenges.

12.6 Development-Induced Displacement

In many countries, including India, the constitution guarantees the right to live in safety and dignity (Article 21). However, large-scale development projects—such as dams, highways, airports, and mining—often disrupt people's livelihoods, undermine their fundamental rights, and lead to forms of forced migration (Vaswani 1992). Each year, an estimated 15 million people are displaced from their homes to make way for such projects, and even higher numbers are affected across Asia and Africa (Aboda et al. 2019; Akinci and Yildirim 2025). This mass displacement gives rise to a wide range of socio-ecological challenges, including homelessness, social fragmentation, marginalisation, loss of access to shared resources, unemployment, food and health insecurity, and landlessness.

Eco-social workers can play a transformative role in supporting affected communities by addressing both the social and environmental dimensions of displacement. Their contributions span micro, mezzo, and macro levels, encompassing activities such as advocating for the rights of displaced people, facilitating Social Impact Assessments (SIA), supporting livelihood restoration, providing psychosocial assistance, rebuilding social networks, fostering cultural and spiritual renewal, conducting participatory risk assessments, and monitoring and evaluating post-resettlement well-being.

12.7 Sustainable Waste Management

One of the primary challenges facing the world currently is waste generation, an unavoidable consequence of human activity. It includes a broad range of things, including industrial wastes and household trash. Waste must be managed effectively to reduce its negative effects on the environment and human health. Waste management concerns multiple aspects of society in addition to the environmental impacts, and it extends beyond mere technical solutions (Urooj et al. 2024). Eco-social workers need to understand waste management as a complex system. Eco-social work, as an integrated field of practice that tackles the nexus of social justice, community well-being, and environmental sustainability, is a critical contributor to sustainable waste management.

Among the many vital eco-social work strategies for sustainable waste management are facilitating sustainable behaviour change, empowering local communities, and strengthening downstream waste management processes such as collection, transport, and treatment. These efforts are complemented by promoting eco-technical interventions—such as composting, bio-treatment, waste-to-energy initiatives, and material recovery facilities—and by engaging in participatory research with waste pickers and community members to understand their needs, capacities, and perspectives. Eco-social work also supports the exploration of circular economy approaches, community-led composting projects, creative reuse initiatives, and stronger recognition and support for informal waste workers. Beyond the direct community interventions, eco-social workers advocate for equitable policies like the Polluter Pays Principle (PPP) or the Extended Producer Responsibility (EPR), which holds primary producers accountable for the entire lifecycle of their products and subsequent waste generation (Lutringer 2017).

Non-governmental organizations (NGOs) play a vital role in sustainable waste management through an eco-social work perspective. By encouraging grassroots environmental action, spreading awareness, pushing for legislative changes, and enabling marginalised people to confront the social effects of environmental degradation, they serve as crucial middlemen. By planning cleanup programs, awareness campaigns, and social action projects that encourage appropriate garbage disposal and community involvement, NGOs put eco-social work into action. Hasiru Dala (https://hasirudala.in/), operating in the Bangalore region of Karnataka state, India, and Green Worms (https://www.greenworms.org/), operating all over Kerala state, India, are two excellent examples of NGO interventions in sustainable waste management.

12.8 Green Disaster Recovery

Disaster recovery is an under-recognized component of emergency management when compared to other phases like mitigation, preparedness and response (Smith and Wenger 2007). UN recommends Build Back Better approach for disaster recovery (UNDRR 2017), which is a practice that reduces risk and vulnerability for both people and nature and supports long-term sustainability. In the context of disaster recovery and reconstruction, eco-social work might offer a vital connection between environmental sustainability and social justice. Following the adoption of the New Eco-Social World, the International Federation of Social Workers (2022) provides a framework for applying eco-social ideas into practice, including disaster recovery and reconstruction efforts. Disaster recovery efforts need a balance between human development and environmental preservation; only then will the following generations be able to sustain it. More research and interventions are required from eco-social work practitioners in energy-efficient, affordable housing, sustainable local food systems, promoting circular economy principles that minimize waste, maximize resource efficiency in rebuilding efforts, and incorporate indigenous knowledge and perspectives in disaster recovery. Sustainable disaster recovery is essential to build long-term resilience and lower future disaster vulnerabilities.

12.9 Humanitarian Interventions and Complex Emergencies

Contemporary humanitarian emergencies are becoming increasingly complex and have multiple triggers. Large-scale economic and social devastation, as well as mass displacement, strained institutions, and economic and societal destruction, are characteristics of complex situations (Paul et al. 2024). Armed conflicts and wars, natural disasters, etc., cause forced migration and environmental degradation. So, various socio-ecological challenges associated with these complex emergencies, such as the unsustainable use of natural resources, inadequate waste disposal-associated disease, habitat destruction and loss of biodiversity, refugee crisis and livelihood insecurities, gender-based violence and exploitation, and breakdown of social services, will be a growing concern for eco-social workers. Eco-social workers often work with humanitarian agencies to deliver psychological first aid and livelihood support. Ensuring WASH (Water supply, Sanitation and Hygiene promotion) standards is another sector that needs to ensure social and environmental justice. Eco-social workers can focus on psychosocial and environmental healing (e.g., ecotherapy) and sustainable rehabilitation, raise awareness of environmental crimes in complex emergencies, and promote bottom-up solutions with environmental stewardship.

12.10 Education and Research

Addressing the complex realities of climate change and related challenges requires the integration of eco-social work into both education and research. Dominelli (2014) notes that social work curricula are already overcrowded, which has limited the space available for disaster management and eco-social perspectives. Yet, eco-social work expands the competencies of future practitioners by incorporating social, ecological, and economic sustainability dimensions (Wu 2021; Tadesse and Obeng 2023; Kang et al. 2025). Integrating eco-social work into social work curricula should therefore not be viewed as a specialized add-on, but rather as a necessary evolution of the profession.

To achieve this, academics can develop pedagogical strategies that embed eco-social perspectives into core courses—such as human behaviour in the social environment, climate change legislation, and ecotherapy—supported by meaningful field placement opportunities. This approach calls for a broader reorientation of social work education, where eco-social work is woven seamlessly into existing frameworks rather than treated as a separate subject area.

Insights from the literature suggest multiple pathways for incorporating eco-social work into social work education. Strategic measures include:

- Incorporating eco-social work as a recognized knowledge domain in the curriculum.
- Prioritizing nature-centric courses alongside conventional anthropocentric ones, particularly in planning, policy, and human service organizations.
- Fostering co-learning opportunities with vulnerable groups such as climate refugees.
- Introducing certificate programs and case studies focused on eco-social interventions.
- Embedding experiential learning opportunities that directly address contemporary vulnerabilities.

12.11 Multidisciplinary Collaboration

Eco-social work and disaster risk reduction address complex societal issues like climate change and social vulnerability. It is essential to connect concepts and methods from different fields. Eco-social work integrates ecological, environmental and social justice and disaster risk reduction, also addressing complex, interconnected issues. The Sendai Framework for Disaster Risk Reduction 2015–2030 clearly identifies the importance of an inclusive multi-stakeholder approach that engages many stakeholders across sectors and disciplines. Interdisciplinary education programs represent a crucial foundation for this transdisciplinary celebration. For example, San Diego State University has developed specialized programs such as their Environmental Social Work and Community Engagement specialization. In India, Loyola College of Social Sciences has a Master of Social Work in Disaster Management. Additionally, Loyola College's MSW Disaster Management program has a technological specialization that allows students to use knowledge from academic topics, including geography, remote sensing, and geographical information systems.

The following are some possible academic areas for productive collaboration in disaster risk reduction and eco-social work: (i) environmental science, (ii) geography/geoscience, (iii) public health, (iv) law/legal studies, (v) economics/development studies, (vi) anthropology/sociology, (vii) psychology/mental health, (viii) political science (ix) public policy, (x) information technology and data science, and (xi) agriculture and food security. These fields can make a substantial contribution to the academic integration of DRR and eco-social work. Effective eco-social work in disaster risk reduction requires multidisciplinary collaboration, which is not merely desirable but mandatory.

12.12 Conclusion

Eco-social work emerges not merely as a subfield but as a metamorphic and essential paradigm shift within the social work profession that addresses contemporary socio-ecological problems, particularly within the domain of DRR. Moving beyond the traditional anthropocentric angles, the eco-social work paradigm embraces an eco-centric approach that situates human well-being within the broader health of the planet. This shift aligns with the International Federation of Social Workers (IFSW) call for a new "eco-social world," which emphasizes the integration of people, planet, prosperity, peace, and partnership. Furthermore, the eco-social work approach serves as a vital connective domain, linking professional social work practice to overarching global agendas like the Sendai Framework for Disaster Risk Reduction (2015–2030) and the Sustainable Development Goals (SDGs). The identified emerging areas underscore the potential for future research to further operationalize eco-social work within disaster contexts, strengthening its relevance as both a practice framework and a research domain. Themes that range from the Anthropocene to interdisciplinary collaboration present many possibilities for further study, practice innovation, and capacity building in eco-social work and DRR, ultimately assisting in the development of more resilient, just, and sustainable societies.

References

Aboda C, Mugagga F, Byakagaba P, Nabanoga G (2019) Development induced displacement; a review of risks faced by communities in developing countries. Soc Anthropol 7(2):100–110. https://doi.org/10.13189/sa.2019.070205

Akinci M, Yildirim F (2025) Mining-induced environmental degradation and displacement in the context of ecosocial work: a qualitative study in rural areas. Sustainability 17(17):7757. https://doi.org/10.3390/su17177757

Berger RM, Kelly JJ (1993) Social work in the ecological crisis. Soc Work 38(5):521–526. https://doi.org/10.1093/sw/38.5.521

Breen K, Greig M, Wu H (2023) Learning green social work in global disaster contexts: a case study approach. Sociol Sci 12(5):288. https://doi.org/10.3390/socsci12050288

Chang E, Sjöberg S, Turunen P, Rambaree K (2025) A call for ecosocial community work: challenges and possibilities for ecosocial work in local neighbourhoods in Sweden. Eur J Soc Work 1–17. https://doi.org/10.1080/13691457.2025.2510340

Coates J (2003) Ecology and social work: toward a new paradigm. Fernwood Publishing, Halifax

Cordoba PS, Bando L (2021) Social work and the sustainable development goals: a student placement model. Aust Soc Work 75(4):519–526. https://doi.org/10.1080/0312407X.2021.1947335

Corlett RT (2015) The anthropocene concept in ecology and conservation. Trends Ecol Evol 30(1):36–41. https://doi.org/10.1016/j.tree.2014.10.007

Dick J, Carruthers-Jones J, Carver S, Miller JD, Dobel AJ, Garbutt A, Hester A, Hails R et al (2020) How are nature-based solutions contributing to priority societal challenges surrounding human well-being in the United Kingdom: a systematic map. Environ Evid 9(1):25. https://doi.org/10.1186/s13750-020-00208-6

Dominelli L (2014) Learning from our past: climate change and disaster interventions in practice. In: Noble C, Strauss H, Littlechild B (eds) Global social work: crossing borders, blurring boundaries. Sydney University Press, Sydney

Gray M, Coates J (2012) Environmental ethics for social work: social work's responsibility to the non-human world. Int J Soc Welf 21:239–247. https://doi.org/10.1111/j.1468-2397.2011.00852.x

Hackmann H, Moser SC, St. Clair AL (2014) The social heart of global environmental change. Nat Clim Chang 4(9):653–655. https://doi.org/10.1038/nclimate2320

Hoffmann-Riem H, Biber-Klemm S, Grossenbacher-Mansuy W, Hirsch Hadorn G, Joye D, Pohl C, Wiesmann U (2008) Idea of the handbook. In: Hirsch Hadorn G et al (eds) Handbook of transdisciplinary research. Springer, Berlin, pp 19–39

Hussain S, Miraj S, Saddique R (2019) Social work and community-based disaster risk management in Pakistan. Pakistan J Appl Soc Sci 10(1):105. https://doi.org/10.46568/pjass.v10i1.105

International Federation of Social Workers (2022) The role of social workers in advancing a new eco-social world. https://www.ifsw.org/the-role-of-social-workers-in-advancing-a-new-eco-social-world

Kang J, Park TK, Yi EHG, Peper M (2025) Ecosocial work and social work education: a scoping review of literature on teaching and curriculum development in ecosocial work. Soc Work Educ:1–23. https://doi.org/10.1080/02615479.2025.2555576

Lutringer C (2017) Environmental and social justice in solid waste management. Policy Develop 10:148–161. https://journals.openedition.org/poldev/2487

Nyahunda L, Matlakala FK, Makhubele JC (2024) Situating social work methods in climate change interventions in the south African context. J Hum Rights Soc Work 9:555–563. https://doi.org/10.1007/s41134-024-00301-7

Panagiotaros CV, Boddy J, Gray T, Ife J (2022) (Re-)imagining social work in the anthropocene. Br J Soc Work 52(8):4778–4794. https://doi.org/10.1093/bjsw/bcac075

Paul K, Denov M, Al Houssami L (2024) Social work practice in complex emergencies: a study of Northwest Syria. J Soc Work 25(3):318–337. https://doi.org/10.1177/14680173241283389

Rice K, Fisher C, Moore S (2022) Social work and the sustainable development goals: an introduction to the special issue. Soc Work Educ 41(7):1409–1411. https://doi.org/10.1080/02615479.2022.2122536

Schmitz CL, Matyók T, Sloan LM, James C (2012) The relationship between social work and environmental sustainability: implications for interdisciplinary practice. Int J Soc Welf 21(3):278–286. https://doi.org/10.1111/j.1468-2397.2011.00855.x

Smith GP, Wenger D (2007) Sustainable disaster recovery: operationalizing an existing agenda. In: Handbook of disaster research. Handbooks of sociology and social research. Springer, New York, NY. https://doi.org/10.1007/978-0-387-32353-4_14

Sofia RD, Abraham A, Little Flower FXL (2024) A review on social work and climate change: the need for a transdisciplinary approach. J Soc Work Pract 39(2):149–163. https://doi.org/10.1080/02650533.2024.2411048

Tadesse ME, Obeng J (2023) An ecosocial work model for social work education in Africa. Afr J Soc Work 13:57–69. https://doi.org/10.4314/ajsw.v13i2.2

United Nations Office for Disaster Risk Reduction (2017) Build back better. The Sendai Framework Terminology on Disaster Risk Reduction. https://www.undrr.org/terminology/build-back-better

Urooj R, Tariq N, Sajjad N, Zaheer N (2024) Socio-environmental impacts of solid waste management. Kurdish Stud 12:78–91. https://doi.org/10.53555/ks.v12i4.2940

Vaswani K (1992) Rehabilitation laws and policies: a critical look. In: Thukral EG (ed) Big dams, displaced people: rivers of sorrow, rivers of change. Sage Publications, New Delhi, pp 155–168

Wellmann T, Andersson E, Knapp S, Lausch A, Palliwoda J, Priess J, Scheuer S, Haase D (2023) Reinforcing nature-based solutions through tools providing social-ecological-technological integration. Ambio 52(2):489–507. https://doi.org/10.1007/s13280-022-01801-4

Wu H (2021) Integration of the disaster component into social work curriculum: teaching undergraduate social work research methods course during COVID-19. Br J Soc Work 51:1799–1819

GPSR Compliance

The European Union's (EU) General Product Safety Regulation (GPSR) is a set of rules that requires consumer products to be safe and our obligations to ensure this.

If you have any concerns about our products, you can contact us on ProductSafety@springernature.com

In case Publisher is established outside the EU, the EU authorized representative is:

Springer Nature Customer Service Center GmbH
Europaplatz 3
69115 Heidelberg, Germany

Batch number: 10777861

Printed by Printforce, the Netherlands